This book is not an official book and has not been approved by the brands that are mentioned. It was created for information purposes only and is intended to help you recognize counterfeits.

Translation:
Susan Jacquet, Transcribe (Timely English)

Photo credits:
Eric Hamdi (MMC – Montres Modernes et de collection), Stéfano Mazzariol, Premiumwatch, Atelier du Bracelet Parisien.

Special thanks to:
Michel Arnoux and the teams of the Federation of the Swiss Watch Industry for their valuable help and for providing the seized counterfeit models used to illustrate this book.
Stephan Ciejka, a key watch industry figure who has supported me right from my early days in this complex and fascinating world.
The Fondation de la Haute Horlogerie for its tireless contribution to the fight against counterfeiting.

ISBN: 978-940506-24-8

Cover page:
left-hand side of the watch: original, right-hand side of the watch: fake

Fabrice Guéroux

watches

An Identification Manual for Contemporary and Collector's Pieces

watchprint.com
publications on jewellery and watches

Contents

Fabrice Guéroux

Specialist in horological counterfeiting and luxury watchmaking

Fabrice Guéroux is an authority on horological counterfeiting and the author of two books about the subject, *Vraies et fausses montres* (real and fake watches).

He advises, among others, the French customs service in lectures and workshops on how to detect horological fakes. Every year he holds master classes at the Geneva High School of Management with the Federation of the Swiss Watch Industry. Many watch companies call on him for teaching or advice relating to counterfeiting.

Backed by more than a decade of experience in identifying fake timepieces, his training courses and the awareness campaigns conducted through his Truetime association make an important contribution to the struggle against horological counterfeiting around the world.

Fabrice Guéroux took part in the *Capital* program on the M6 TV channel investigating fake Rolex watch distribution networks and published articles in specialized media, including *La Revue des Montres*, *Heures*, and *Night & Day*. The French press — *Le Parisien*, *Le Nouvel Economiste*, *Montres Magazine*, *Ouest France* — has covered his activities extensively.

Fabrice Guéroux currently manages two Internet sites about watchmaking:
www.fabricegueroux.com (horological news)
www.truetime.fr (watch counterfeiting news)

Fabrice Guéroux
watchmanual@gmail.com

*This manual will help you learn to appreciate the care taken
in finishing the tiniest details of a properly manufactured watch.
By examining and comparing the level of finish, you can tell
the difference between a fake and a genuine watch.
In all modesty I hereby make available to you the fruit of several
years of observation.*

Fabrice Guéroux

How to use this book

This book is a sequel to two volumes of *Vraies et fausses montres* (real and fake watches). They were written to enable anyone to improve their ability to tell the difference between a fake and a genuine watch.
Published in 2007, the first volume focused on photographs comparing real and fake watches. Four years later, a second collection included new watch models in the market for counterfeits.
This manual projects a realistic way in which a watch should be examined for identification or authentication, in view of the rapid changes in the quality of counterfeit products in recent years. It has become extremely difficult these days to examine some watch models and make a judgment on their authenticity without a minute analysis of their details and features. Previously, the quality of the materials, the accuracy of the markings, the proportions, weight and other factors made for relatively easy appraisals. Now these are sadly the features that counterfeiters are getting better at copying.
I predicted this development a few years ago. I knew that the revenues generated by the sale of fakes online would give rise to a true parallel market. This state of affairs would enable counterfeiters to invest in ways of improving the quality of their products. Fraudulent gains have thus enabled this industry to invest in the development of advanced technological tools while acquiring machine tools that are identical to those used by some watch manufacturers. As a result, production units equipped with good-quality machinery have been established to make ever more exact copies of each technical development by the watch brands.
You now have to sharpen your observation skills in order to appreciate the finishes of a vintage or contemporary watch. It's a task akin to that of an art expert who focuses on the important details of a work to distinguish coherent features that might lead him to certify the work as an original. It's a thrilling job – an investigation that will reveal the characteristics of legitimately manufactured watch models. At the same time, you will be taken into the micromechanical world and the horological finishes that give fine watchmaking its current status.
This third volume finds its worth in the investigations and research that I have undertaken. Whenever a new counterfeit model came to light, by the time my examination revealed how the counterfeiters had produced it, a new and more sophisticated model appeared on the market for fakes and my analytical task had to start all over again.
The main purpose of this manual is to enable the reader to make similar discoveries to those I have made during the past few years. To this end, I intended to make publicly available a large part of the comparative features in my possession, initially as images then as detailed descriptions of my analyses. Three watches were literally dissected and autopsied in a richly illustrated chapter. They will enable you to ap-

preciate the differences in manufacturing quality between an original watch and a fake. Even if a fake looks authentic, the truth can always be revealed by showing certain details.
An entire chapter is devoted to the vintage watch and its fakes. It provides a useful tool for all watch enthusiasts.
To conclude, the best response by the brands to the scourge of counterfeiting is continuous investment in the quality of manufacturing and high-grade finishes. It stands to reason that the money spent by watch manufacturers in marketing, advertising, legal services, boutiques and the fees of their ambassadors cannot be compared with any outlay by counterfeiters who indirectly benefit from the costs borne by the watch companies. The counterfeiter only has to produce and sell. If you learn to tell the difference between a perfectly machined component and a similar part that is slightly less than perfect, you will discover two worlds: that of fine watchmaking and that of horological fakery.
It is with this perspective that you can use this book and I earnestly hope that it will tell you everything you want to know about these two subjects.

Enjoy reading!

Introduction

After the publication of my first two books I continued to investigate more closely the subject of counterfeiting. It allows me to understand the trends and development of this phenomenon for the next 10 years.

For now more than 10 years, the counterfeiting phenomenon has done nothing but expand. It's a source of worry to the industries affected by this illegal market of fake medication, components and luxury products. These and other counterfeit products often make headlines and poison the economic system.

What are the consequences of counterfeiting for today's economy? This parallel industry affects the economic framework as well as the safety of consumers. The opportunity costs for watch companies seem obvious and the legal consequences of purchasing such goods are starting to gain momentum.

It must be understood that the counterfeit market is opaque and is linked to organized crime. Making the public aware of the counterfeiting problem is not futile and is beginning to have a real impact on buying behavior.

Counterfeiting watches is growing exponentially, especially concerning online sales. Judicial authorities struggle and take recurrent action against illegal sales sites while there has been a resurgence of counterfeited bands. This phenomenon is no longer confined to the major brands but affects the entire luxury watch sector. The result is a constant increase in the choice and range of counterfeit models.

From an economics point of view, one could say that on the face of it counterfeiting cannot have any real impact on the companies because they always have the means to defend themselves. However, the expansion of this parallel economy has a butterfly effect on the watchmaking industry and we are beginning to see its first damaging repercussions.

In 2006, my first book laid down the premises of an invasion of fakes and already foresaw a resurgence of sales in view of the endless possibilities of the Internet. The counterfeiters already had a suitable terrain as well as the facilities with which to spread their fake watches. From then on, copies with a modicum of "quality" were being noticed, but failed to get much attention from watch connoisseurs, experts and other enthusiasts.

In 2010 it became necessary to update some of the data and evaluations in a second book. The counterfeiters had in fact made a great leap forward in the apparent quality of their copies. As a result, the quantities disposed of saw a considerable rise. The brands, which had hitherto little reason to be concerned, started taking a closer interest in the issue.

The following years allowed me to dig deeper into the field of counterfeiting and I endeavored to assess more precisely the directions and trends of this phenomenon in the next 10 years. In this volume I cover the subject of vintage watches and devote much of this manual to the subject.

It's because of this expertise that quite a number of my readers send me emails and call me for information on a period model that they're interested in. I'm always happy to help and ensure that an enthusiast makes a purchase in the right conditions.
The type of requests and questions that I get enable me to confirm today that the vintage-watch market is literally flooded with fakes. Fitting a watch with fake parts, such as a clasp or a crown, can increase its value tenfold. I notice that quite a few watch fans are taking an interest in the vintage watch market, considering it a road to rich returns. I should also point out that counterfeiters know how to read, they are well organized and can follow the developments and trends of a market. Furthermore, they have the capacity to meet any given demand, often faster than some professionals. I can only conclude that the surprises are not over.
So far as contemporary watches are concerned, I thought that the "quality" of the common copies would have improved, as in the 1980s which saw a resurgence of fake Rolexes, Cartiers, Breitlings and others that could be found in street markets. Nothing has changed on that side. A recent hidden-camera documentary showed that such markets were still the playground of the small retailers of low-grade fakes. The way of presenting the product is always the same, and when you ask the seller for something better you are quickly taken to a side alley. There a new seller, introducing himself as an "expert supplier of rare objects", will offer watches of superior quality, sold for between 500 and 1,000 euros. This buys you the entry level in the hierarchy of quality fakes.
Openly announced as counterfeit products, such watches can sometimes fetch 2,500 euros. On the same subject, two factors have drawn my attention. The first is the number of brands featured in the counterfeiters' catalogs. The second is the number of fake complicated watches on offer.
In 2013, my view of the market in fakes could be summed up in the equation: top brand + popular model + quality product = market in fakes, consisting predominantly of low-grade copies. I was wrong.
The priority has definitely been the effort put into the two factors mentioned above. Judging from what I have found out, even independent watchmakers will soon have to recruit a team of lawyers specialized in intellectual property rights and the suppression of fakes. The niche brands are in the counterfeiters' catalogs as well, and even though the quality is not always there, examples can be found that could fool quite a few. Counterfeiters are undeniably a danger for the watch industry. Most of them have no problem in declaring that the products they are selling are fake. At this stage the buyer knows what he's in for and his purchase is his own responsibility. The person-to-person sales sites for their part are overflowing with copies under cover of some kind of anonymity for sellers who claim to sell authentic watches.
Having read so far you will probably end up by thinking that it is difficult, if not impossible to get a good deal without the company or the advice of an expert, and that it is dangerous to commit financially to such a sensitive issue. I would say that learning and understanding are important. Education in such an exacting subject as horology requires some learning and knowledge. That is what I shall try to bring you in this book.

Fabrice Guéroux

Fake Hublot watch. Functional tourbillon mechanism.

Understanding the counterfeiting industry

It is impossible to put a figure on the number of counterfeit horological products that have been sold in the past few years. To do so would require an estimate of the number of units actually produced, but unfortunately making fakes is an activity over which there is not much control. Nevertheless, there are indications that the 2005 production figure of around 40 million fakes a year has been steadily increasing since.
Guesses as to the exact number of fake watches produced a year and revenues from counterfeiting vary somewhat according to the sources able to provide information on the subject. The figures are thus fairly random. By putting together various facts it is nevertheless possible to see that the increase in production is real. Some authorities, notably the Federation of the Swiss Watch Industry (FH), seize increasing numbers of fake watches. This tells us that the market is clearly making net progress. For example, the investigative branch of the FH has set up coordination with the authorities in Dubai. A total of 90,000 fakes were confiscated in the city in 2013 alone. Add the fakes that got away and multiply that by the number of countries and cities that have become notorious as counterfeit distribution centers, and it is easy to conclude that this is a flourishing market where tens of millions of fakes are disposed of.
However, assessing these figures and calculating others is not the purpose of this book. The profit shortfall occasioned by this fraudulent trade is mainly a matter for the brands. As for the direct consumers, they are far more interested in finding out how to avoid having fake watches on their wrists.

Assets deployed by counterfeiters

The production of fakes raises a swirl of questions. I often get requests from Internet users, students, journalists or various watch enthusiasts who want to know where the factories making all these counterfeit watches are hidden. Some are led to believe that such places exist. It is evidently a hidden production system, secretive enough not to have their signs hanging out on the high street.
The best way of tackling the problem is from the production side.
Let's start with the production of original watches in a manufacturing company in which all or some of the production is integrated. The components are designed and produced by the company. As a result, it controls virtually all the production on its premises. It is important to remember that not many watch companies are capable of this.
Most watch producing companies commission sub-contractors for a good number of the production-line operations. These contractors, specialized in making watchcases, dials, hands and even entire movements, supply the parts so that the watch company can complete the assembly of the watch and undertake various quality checks.

These suppliers produce components of excellent quality according to the strict criteria that meet the demands of luxury watchmaking. After all, your timepiece consists of the sum of its parts, each immaculately finished.
The standard demanded by the brand makes the difference. If a watch is to be sold for tens of thousands of euros, the movement, case and dial have to be of commensurate quality. If the consumer is paying less than 1,000 euros for a watch, its quality is necessarily inferior. It is therefore thanks to the skills and excellent workmanship of these suppliers that the specifications set by the watch manufacturers can be met.
The same applies to the production of fake watches. A factory in Asia producing millions of watchcases a year for hundreds of clients, will accept any order provided the specifications are clear and the bills are paid. It will not spend a moment thinking about whether there is any reason to consider some trifling issue involving intellectual property. By that I mean there will be no investigation by any legal department into the ownership or the patents of the components to be produced. Nor will the factory study the features of every branded watch to find out whether its client's manufacturing blueprints have not already been patented. It will produce the case and do so in quantity, making tens of thousands of copies for the client. The same applies to the entire production line.
Let's now turn to China and the production of horological parts coming out of that country. Take the example of the production line of a fake watch. The original watch is fitted with a case in high-grade steel. The counterfeiter will order an identical case and do the same for all the constituent parts of the original watch. So far as the movement is concerned, let us suppose that the original watch is fitted with an ETA selfwinding movement. It too will be taken apart and its constituent parts copied.
It is only when all the parts needed to assemble a watch have been produced that the counterfeiter can embark on stage one of his business.
He will therefore have the watches assembled and will concentrate on making them identical to the original, going so far as to copy the brand's hallmarks, screen printing and engravings while adding false serial numbers. From this moment on he is acting illegally with intent to commit a straightforward fraud.
In order to dismiss any stereotypes, it must be said that it would be unthinkable for a Chinese businessman with a registered business wishing to establish himself in the watchmaking industry to have in his business plan the creation of an Internet site for the sale and distribution of fake watches.
Let's take the watchcase for example. If you are the lucky owner of a Swiss Made watch, you will notice that the serial number will certainly be on the case. On a Rolex watch it is between the lugs of the case or on the flange in the latest models. On the watches of other makers, the number is engraved on the back of the watch or on the screw caseback. You will notice that in most of the watches the serial number is properly cut into the metal with a resulting relief denoting a quality marking.
This stage is more difficult for counterfeiters because it needs qualified and precise workmanship, while it is often carried out by clandestine workers by hand. As a result, I found fake Rolexes engraved between the lugs with a spelling mistake: " REGISTED" instead of "REGISTERED".
As for the engravings on the back of the watch, such as the serial number or the number of a limited-edition watch, you will notice that they are very often not cut into the metal, but simply inscribed with a laser. It looks as if the engraving is on the surface rather than in depth. The same goes for the stamping of the dial. In an original watch, a lens reveals that it is correctly made with enough relief to avoid deformed letters.

Unmarked fake components awaiting marking.

It is quite different on dials made for counterfeit watches.
However, we should not be too dismissive on these points because we could, in the future, be unpleasantly surprised by the improvement in quality and the determination of the counterfeiters to persist in their fraudulent activity. They constitute a competitive element that now takes a substantial share of the market. The fact that such horological complications as the tourbillon or perpetual calendars are emerging onto this parallel market, leads to the conclusion that their determination is not running out of steam.

What kind of watches are copied today?

To get an idea of the range of watches that are counterfeited today, you have to take a look of the overall economic development of the watch market in the past few years.
Watch counterfeiting was already proving to be an issue before the 2000s and the two or three flagship brands of the Swiss watch industry were beginning to feel the adverse effects of the phenomenon. Although the revenues generated by the counterfeiters were far from today's figures, it should be pointed out that various tourist markets and flea markets had already been invaded by people selling fake watches.
In those days it was already worth going to the markets in Morocco, Hong Kong or even Ventimiglia which were the cradles of counterfeit distribution.
To give you an example, I remember going to the flea market about 20 years ago to investigate the subject. It was no less than two hours before a seller took from his coat a watch that from a distance might look like a Rolex Datejust in gold and steel. The price asked for this copy was as shocking as its total lack of quality. For the equivalent of 200 euros, the salesman was asking me to blacken my wrists with a corrosive watch. It was not at all convincing.

Fake vintage Rolex Daytona Ref. 6263.

Fake Glashütte Original tourbillon.

Unfortunately, two factors played into the hands of the counterfeiters in the ensuing years. The first is the development of the Internet, which added to open borders, has complicated the task of the customs services. The second is the expansion of the Swiss watch industry both in terms of the turnover from imported products and of the varieties of brands that were hardly talked about until now.
In 2001, nearly 30 million watches were produced in Switzerland for a turnover of around 10 billion Swiss francs. The figures stayed about the same with a drop in 2003. Another billion in turnover broke a record in 2004. Sales grew by another billion in 2005 to set a new Swiss industry record. The following year saw a surge in demand for luxury products. It didn't stop there, because in 2007 sales grew by another 2.2 billion.
Note that revenues for 2012 amounted to 21.4 billion francs and these results have remained constant since. It is clear that the demand for Swiss manufactured products has not ceased to grow, giving rise to what we are witnessing today: the return of many brands that are today classified as luxury watches.
Aided by the Internet's communication and trading facilities, counterfeiters have seized this opportunity to change their strategy and that is without counting the growing demand for luxury products and high-grade watchmaking. The growth curves have followed the same trend as the Swiss watch industry, and if you consider that the production of fake products is at least 40% higher than that of the original items, you begin to get an idea of the collateral damage.
Ten years ago, we could observe that the counterfeit market was invaded by fake Rolexes (more than 80%). Today such products as the Audemars Piguet Royal Oak are copied with unparalleled ease. One can clearly state that practically all watch brands are copied.
From 2012 it became apparent that some of the high-grade luxury watch brands were being counterfeited. From time to time I came across a counterfeit model, which although quite easy to identify, was astonishingly well made. One example was the copy of an IWC Portugaise F.A. Jones, with a slightly complicated movement visible through the caseback. Previously, counterfeiters worked only on the visible parts of the watch and the decorative parts of the movement.
During my latest investigations, I wasn't surprised to find fake Patek Philippes but also a massive presence of counterfeit Richard Mille or even Roger Dubuis brands.

Distribution

It's not possible to provide precise sales figures of fake products in this manual. Nevertheless, it is vital to have an idea of the scale of the market for fakes so that we can become aware of the dangers it presents for the second-hand market. Why such a large turnover when the distribution network is hardly visible?
Although it's obvious that a business making fake watches will never undertake marketing campaigns like those of the great watch brands, it should nevertheless be remembered that there are other forms of marketing available. The Internet has today become the market place for the huge traffic in counterfeit products. Over the past few years millions of objects have found buyers and changed hands. These sites are the main target of brands and professional bodies. As soon as an illegal site is closed down, there's nothing easier than to open another under a different name or on another platform. For the counterfeiters, quantities and search-engine referrals are more important than quality. Sites are duplicated and even if you consider a site that is closed down and reopens within a week can go from 100 watches sold per day to 10 a day at the beginning, you have to multiply the number of sites by a few thousand and then add the mass spamming campaigns and effective communication on social networks, forums and second-hand sales sites. You will then find it easy

Fake Audemars Piguet Ref. 15400. While the effect is indeed "impressive", what we see here is merely a decorative facade (see full details for this reference on page 171).

to realize the extent of this phenomenon and the damage it does to the watch industry.
If this market didn't benefit from the activities of organized crime, in other words the thousands of individuals exploited for sales, production, assembly and the setting up of Internet sites, things would be very different today. The counterfeit market can be compared to drug trafficking, which has long been impossible to root out. Millions of fake objects circulate today, while the number of consumers of all ages has grown impressively.
It is not unusual to see these watches arriving at the after-sales department of the brands or at independent watchmakers. While a great number of these fail at the first checkpoint and are destined for immediate destruction, some manage to escape the vigilance of unaware watch repairers unused to having fakes of superior quality in their hands. Such watches, of course, continue their journey and very often end up on the second-hand market.
While fake watches were initially sold as such, we are coming across increasing numbers of sales by individuals who are not aware of the kind if watch that they are trying to sell on the second-hand market and often remain convinced that it's an original watch.

UNITS PER
ROLEX
OYSTER
COSMOGRAPH
DAYTONA
SWISS

Previously owned watches and vintage fakes

The second-hand and vintage watch market has seen its structure change somewhat in the past 15 years. In the 1990s, there were only a few dealers in the market, but it's not like that today. That can be easily explained by looking at the price differences between 1990 and today of most watch models. The vintage trade has now emerged from its niche to take part in the watch exchanges and the major auction houses. Record horological sales are not unusual.

Public fascination is the obvious reason for this economic turnaround in vintage and contemporary watchmaking. Generalist economists or specialists in this sector give various logical and rational reasons for this phenomenon.

Watches that used to be considered expensive 15 years ago have in fact today moved into the category of the unaffordable and have simply become full members of the pantheon of collectors' watches.

Rather than tackle the speculative end of this phenomenon, it would be more profitable to look at the revolution it has triggered in the domain that is of such interest to us.

Even if today some people still think that the markets for previously owned watches and for new watches are separate and different, there are obvious clues that point to the opposite state of affairs. To support this theory, we can take a few collector's items that have seen their value rise and look at how this has affected the market for new watches.

If you consult the auction catalogs, you will notice that the run-of-the-mill lots have today very often turned into sought-after items. Take for example such models as the Rolex Submariner or the GMT-Master. These watches sell very well and are among the brand's flagship models. They have become best sellers that are sometimes sold for stratospheric prices. Getting a bargain is virtually out of the question; instead you get a fixed quote. Added to this phenomenon are the so-called arbitrary watches, fabricated models like the "red" version of the Submariner (the initial production of the reference 1680) or the GMT-Master "Cornino" of the 1960s with pointed lugs protecting the crown. Of course, the "Cornino" is not as common as the ordinary 1675, but it seems rather excessive to see a model that has sold in its thousands treble in price especially if it is in poor condition.

Blaming collectors or decrying the fabrication of rare models is not the intention here. If we keep strictly to the subject of this book, we come to realize that he door is opening ever wider to counterfeit products. If a small detail of a mass-produced timepiece can make such a difference, then it should be possible and very easy for a counterfeiter to produce a copy and treble his stake. For example, what can be done with an ordinary Submariner reference 1680? It's a model that is priced between 5,000 and 6,000 euros. But supposing you turned it into a "Red Submariner", a "Comex", or even into

Opposite page: fake vintage Rolex Daytona Ref. 6263.

a model made to order for the Sultan of Oman? You can appreciate that the possibilities are unlimited.
The person who buys this kind of model is nearly always trying to turn a profit. He would have noticed that its price doubled between 2005 and 2014. But what makes such a watch so highly prized? It could be its dial, for example. But there's nothing easier than to forge a dial. And what about the buyer who within five years finds out that his Rolex Red Submariner simply a fake? If he wants to give it back its authenticity, he has to find an original dial, preferably one with the right patina that has not been repainted. This type of dial today changes hands for about 1,000 euros. Thus, if a collector has to invest such a sum to fit an original dial, he also faces a loss of around 3,000 euros for his watch to regain its authenticity. It's not a particularly alluring outlook, especially since the likelihood of getting back to the vendor is very small.
There are dozens of similar examples and many people pay the price and suffer the consequences. You might recover from being swindled in the purchase of a Red Submariner, but if instead it involves a so-called "Comex" or a Paul Newman Daytona Panda 6262, the buyer has to spend a lot more to find himself the owner of a vintage watch that is worth less than its estimated value.
It is therefore interesting to observe that counterfeiting does not only affect contemporary watches. The vintage models that are sought after in the watch market have seen their prices climb impressively, and to everyone's surprise, the price of some models doesn't look ready to drop. The market for collectors' watches has begun to be tainted by fakes and the situation is getting worse. Contemporary counterfeiters have realized this and have no compunction in putting on the market vintage Rolex models like the "Big Red" (dials with Daytona inscribed in red) fitted with Chinese selfwinding movements — peculiar for a watch that was originally manually wound!
Following the market trends for collectors' watches has become relatively complicated. All it takes to skew the market is for one or other model to reach a record price. A well-known example is the Jean-Claude Killy Rolex (reference 6236) sold for almost 650,000 at Sotheby's in 2013 — rather a lot for a watch that was previously changing hands for 160,000 euros at the most. A few weeks after the sale, it became impossible to find this model, in good condition, for less than 250,000 euros. Regarding the last point, it is worth mentioning that the condition of the watch is a criterion that is also disappearing. Some people are prepared to take a leap in the dark and buy an oxidized watch! But how can you find a bargain in a market where counterfeiting is spreading its roots so quickly and easily? The answer is to change your behavior and avoid rushing headlong into a purchase. It's better to think about learning what has today become a new profession.
Moreover, since a watch consists only of a case, a dial, a strap/bracelet, hands and a mechanism, we can say that it's pretty easy to cover all the bases. The most common example one might find would be the following: according to the "experts", it would appear that Rolex produced a series of GMT-Master models for the army. What army? Although there is speculation regarding Peru, Argentina, Saudi Arabia or Italy, there are to date no records or documents enabling us to determine the origin of these models equipped with an all-blue bezel. While it gives the watch a really superb look, especially when the hour markers feature a beautiful orange patina, one naturally also notes that model is rare and therefore very expensive. It is thus possible to find on the market simple Rolex GMT-Master "Blue Bezel" watches selling for between 20,000 and 30,000 euros.
What's the problem, you might say? Well, the bezel is the only factor that makes this 1675 a very rare version. And at the same time, a fake blue bezel can be acquired on the counterfeit market for less than 50 dollars.

Fake Rolex Daytona Ref. 6263, with an automatic movement that was never used to equip these models.

It's difficult under these circumstances to tell the difference. In this instance and in my opinion, the sole means of reliably authenticating the watch would be to sell it with a Rolex repair document containing the watch's reference number followed by the words "Special Order for...".

So that's where the problem lies. Things would be quite different if it were a matter of purely technical details relating to the mechanism or given complication – such as a tourbillon movement for example – but that is definitely not the case here. Moreover, the issue applies as much to current as to vintage watches, such as the fake Big Bang Hublot modern tourbillon model, or the historical Blancpain Fifty Fathoms.

Above: fake Hublot Big Bang Tourbillon (functional).
Opposite page: fake vintage Blancpain Fifty Fathoms.

BLANCPAIN
Fifty Fathoms
19
NO RADIATIONS
SWISS MADE

6-25-00

How to detect a fake

In the different sections of this book I will present watches to compare and to give you the opportunity of examining the original watches and their copies.
In the following pages I will illustrate several vintage and modern examples of fake watches in which we can identify the important details to look at. This is an initial approach.
These details are presented as useful information. Of course, when it comes to a forensic examination, the aim is to collect and classify as many as possible of the differences between watches to make a judgment.
The market for counterfeits is developing so rapidly that many of the hitherto obvious identifying features of fakes tend to disappear.
Some features, often too complicated to copy and in any case absent from the counterfeiter's specification sheet, enable the difference between the real and the fake to be determined quickly. The most obvious is the movement of the watch. This is often the component that is the most difficult to reproduce, especially when it comes to high-grade luxury watchmaking.
Nevertheless, other features are not to be neglected, and typographical mistakes in the markings and inscriptions are multiplying. And that's not counting the spelling mistakes in the reference markings made at the end of the assembly chain by those who cannot speak the language and are not familiar with the alphabet.
The metals or leather used in the bracelets and straps as well as the overall aging of a vintage watch are also aspects to look at. Such details, if taken together, could initially raise our suspicions about a watch and lead to the wise decision not to buy if there are too many inconsistent features.
The watches shown in the following pages were selected with the intention of presenting a range of models each of which has its weak points. It is clear that bringing together all the fakes that are produced nowadays would be impossible.

Opposite page: detail of a fake Hublot BIg Bang folding clasp.

Fake Patek Philippe Nautilus Ref. 5711

Although this fake could mislead certain people, we note that the finishing details are of very poor quality compared with the usual standards upheld by Patek Philippe. The horizontal dial stripes are rounded, instead of straight-angled as they are on the original.

Original Patek Philippe Nautilus Ref. 5711

Fake Patek Philippe Nautilus Ref. 5711

The stem of the hand shaft is machined in 'riveted' mode, whereas on the original the hand shaft is solid and not hollowed.

Patek Philippe logo on the balance wheel of the automatic winding rotor. It is merely glued onto the fake model.

Original Patek Philippe Nautilus Ref. 5711

Patek Philippe logo on the balance wheel of the automatic winding rotor. It is etched into the original.

Fake Patek Philippe Nautilus Ref. 5711

The fake features a poorly machined and beveled case-holding component. On the original, the space between the two parts is extremely narrow and the beveling is far more clear-cut.

Original Patek Philippe Nautilus Ref. 5711

Fake Patek Philippe Nautilus Ref. 5711

The beveling (chamfer) on the fake is twice as large as on the original, on which the chamfer is far more cleanly beveled [1].
The link connecting the case to the bracelet is far too large on the fake [2].

The cut of the case chamfer is very rough.

Original Patek Philippe Nautilus Ref. 5711

Fake Patek Philippe Nautilus Ref. 5712

A non-functional moon phase. When you have a watch in hand, it is important to test its overall operation as well as that of its complication. Also note the different positioning of the complications, mainly the differently sized calendar/moon phase subdial and the absence of an hour marker at 8 o'clock.

Original Patek Philippe Nautilus Ref. 5712

Fake Patek Philippe Nautilus Ref. 5712

The visible mechanism of this fake serves to determine the origin of this watch and its lack of authenticity.
One notes an interesting set of laser engravings aimed at reproducing the maker's marks.
The fake is equipped with an entirely invented manually wound movement, as opposed to the selfwinding micro-rotor movement on the original (right-hand page).

Original Patek Philippe Nautilus Ref. 5712

Fake Rolex Daytona Ref. 6263

Fake Rolex Daytona Ref. 6263

Laser-printed marking on the caseback. This technology did not exist during the years when this model was produced.

Fake Rolex Daytona Ref. 6263

The quality of machining on the crown is worth noting here.

Fake Rolex Daytona Ref. 6263

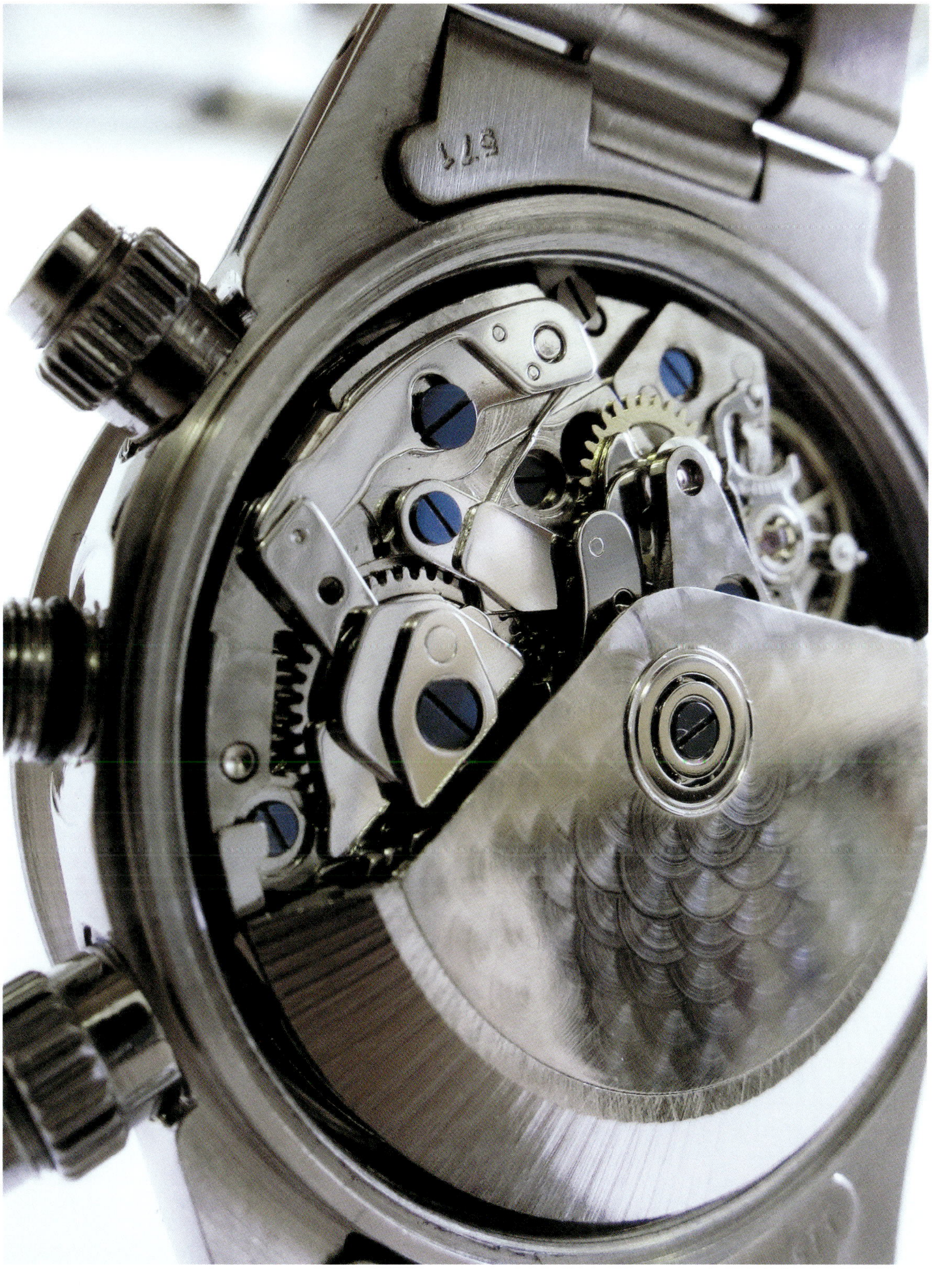

Finally, to conclude this introduction to counterfeiting rare vintage watches, the mechanism will leave no doubt in the mind of most connoisseurs: the original manually wound Rolex movement (a Valjoux caliber modified by Rolex) has been replaced by a Chinese selfwinding movement.

Fake Hublot Big Bang Tourbillon

Complicated movements have not been spared by this scourge and one now sees functional tourbillons fitted on fake luxury watches.

Fake Hublot Big Bang Tourbillon

An interesting factor serving to track down a fake watch: the serial number is the same as on the entire production of this counterfeit. Taking a look at the various counterfeit watch sites and checking their serial numbers for the same model may be an effective method.

The evolution of 'fake' tourbillons in recent years is truly amazing.

Fake Glashütte Original Tourbillon (A. Helwig)

Fake Glashütte Original Tourbillon (A. Helwig)

We note here the extremely poor quality of the brushing on the case. I invite you to examine the Glashütte Original collection and you will see that this model never existed.

Fake Rolex Sea-Dweller Deepsea Ref. 116660

When compared with the original watch, the dial features several font errors. The presence of an opening, as presented on the following page, will enable reliable identification.

Fake Rolex Sea-Dweller Deepsea Ref. 116660

Fake Rolex Sea-Dweller Deepsea Ref. 116660

This fake Sea-Dweller provides a glimpse of a poorly made movement. The counterfeiters have even tried to reproduce the fuchsia pink notched Rolex wheels by sticking two colored dots on a steel wheel. The movement of the original Rolex Sea-Dweller Ref. 116660 is equipped with a free-sprung balance fitted with inertia weights and not with an adjustment index as in this fake. This detail is illustrated on pages 140 and 141.

Fake Rolex Sea-Dweller Deepsea Ref. 116660

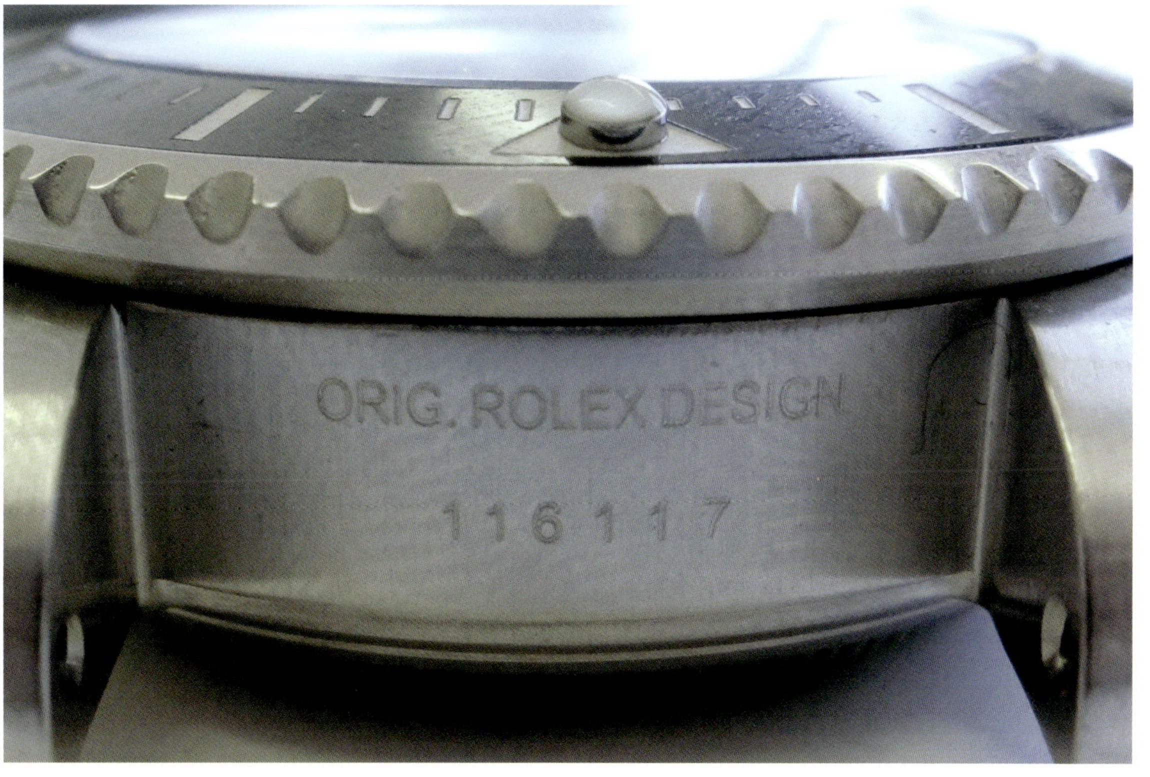

The laser markings are of poor quality, and the reference number of the watch is incorrect (116117 is not a Rolex reference).

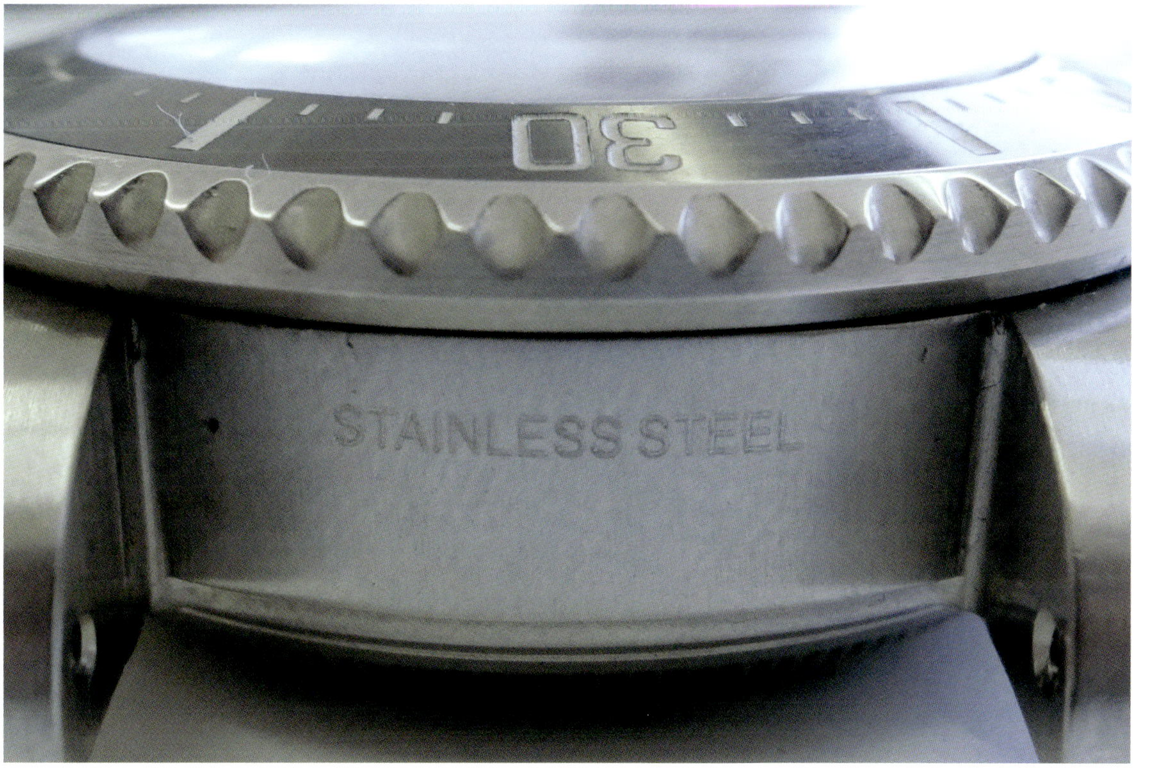

Fake Seamaster Ploprof 1200M Omega Co-Axial

Fake Seamaster Ploprof 1200M Omega Co-Axial

Slight drop on the case not found on the original watch [1]

Highly similar rear view of the watch, apart from a notable difference in terms of the screw-down caseback that is rounded on the fake and straight-cut on the original [2].

Fake Seamaster Ploprof 1200M Omega Co-Axial

While the counterfeiting is extremely well done, the finishes are not entirely satisfactory. The Sea-Master logo on the original [1] is straight cut on each angle, whereas the fake is more rounded (a magnifying glass is a must).

Fake Seamaster Ploprof 1200M Omega Co-Axial

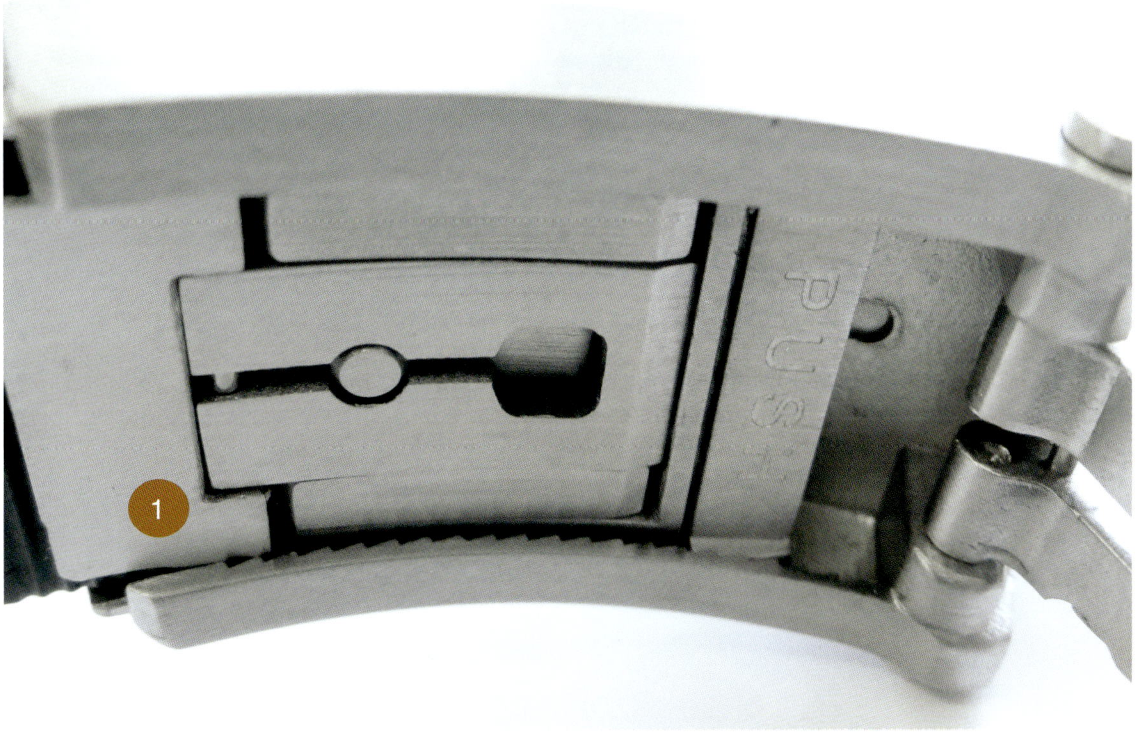

Poorly machined finishes and case on the fake [1] and [2].

TACHYMÈTRE
OMEGA
Speedmaster
PROFESSIONAL

Genuine fake watches

The problem of generic parts

A generic part is a spare part made by a party other than the brand. Unfortunately, it is not uncommon to find on the markets vintage collection watches, being by definition "old", fitted with these kinds of components. They often involve a bridge of the movement, screws, a crown, hands and any other part that has to be repaired or replaced.
Let me tell you a story. One day, someone came to see me saying he was unable to tell the production date of his Omega Speedmaster "Moonwatch" fitted with the caliber 861 movement (pictured left). The movement's serial number had six figures, which was remarkable because it should have had eight. Before he came to see me, he had come up with a host of explanations, the chief of which was that it was a mistake by the brand, which could make his watch very rare. Unfamiliar with his kind of anomaly, I turned to a watchmaker who was specialized in the company concerned. He opened the watch and his verdict was not as encouraging as the dealer had suggested. The bridge had been replaced by a generic part. An impressive deception, given that it was the same color as the other parts. As for the six-figure reference number, we never got to hear the real story. Because of this the watch fell from an estimate of around 3,000 euros to a trifling amount. To bring it back to its previous value, an original part would have to be found.

This example is important, because if at first glance it might seem a trivial incident, what does it say about the parts on the market today? Here the problem is more complicated than for counterfeit watches. It is getting less easy to buy a watch for a bargain price, but it is also essential to consider that the vendors do not spend much time worrying about whether its parts are original or not (this is even more applicable to the movement of the watch). When you try to sell your watch, hopefully for a profit, the bitter taste left by the comment of the potential buyer is difficult to accept when he tells you that he doesn't want to buy because one or two generic parts have to be replaced.
Before you buy a watch, I would advise you to find out about the model: the type of dial the watch been fitted with since it was made, and what movement does it have, backed up by pictures of the movement. For that the Internet and specialist books are today a gold mine.

Forgers

The word "forger" is chosen deliberately. In this regard, I make a clear distinction between forger and counterfeiter. The vast majority of counterfeiters are well established on the web and describe themselves as such. They clearly announce the nature of their business, namely that their site offers "best quality replicas."
The customer in any case goes ahead with the purchase of a counterfeit, and in theory the fraud

Opposite page: Omega Speedmaster "Moonwatch" equipped with fake parts.

only affects the owner of the intellectual property rights, in other words the brand whose product has been copied.
As for the forger, his product is also fake but there's an additional factor: a premeditated attempt to defraud the purchaser. The intention is to deceive the buyer as to the true nature of the product with the aim of making a large profit.
With the soaring sales prices of vintage watches, added to the public obsession with luxury watches, we have seen a real resurgence of counterfeit products entirely or partially fabricated by forgers of varying skills.
For those with less skills, we find on the market their vintage watches made up of original parts but modified to increase their sales price. There are now hundreds of diving or military watches labeled Rolex, Tudor, Omega, Blancpain or Longines that have never been on a military or a diver's wrist. And that is very often apparent. It might look like a time-worn watch, but it is fitted with a freshly painted brand-new dial. You

Fake Rolex military dial (British army) that can be fitted on a Submariner Ref. 5513.

can see these watches on such sites as eBay.com (international) if you search for "Military Omega" or "Military Longines". There are so many timepieces on offer that you would be forgiven for thinking that the army had conducted a fire sale. That is only one example among many. On the same quality level, in Eastern Europe you can find watches labeled Patek Philippe, A. Lange & Söhne or Breguet for less than 3,000 euros.

The common denominator of such relatively cheap watches is their huge diameter (54mm for the smallest). These are in fact pocket-watches on which lugs have been soldered so that a big leather strap can be fitted. The evil is compounded, because they are not satisfied with just offering copies of models that never existed. The forgers base their work on some brand or other's old pocket-watches and then make a pretty enamel dial labeled with the brand that happens to be rising in the current charts. The more meticulous take it upon themselves to engrave the balance cock.

Fake Blancpain Fifty Fathoms "Aqua Lung" watch. Most of the parts have been remade, including the dial, bezel, disk, case, movement and back of the watch.

The more skilful counterfeiters create the trickiest problems. What about the basic Rolex Submariner 5513, worth around 4,500 euros, modified to become a military Submariner reference 5517 that sells for 40,000 euros? Apart from the huge price gap, the only differences between the two models reside in the inscriptions on the back, the dial, the hands and the location of the reference number on one of the lugs (for the military model). This is just one example of a forgery intended to deceive and to fraudulently sell a watch purporting to be an original.

I was recently given rather a peculiar watch to look at. A forger had amused himself by creating a fake vintage Rolex GMT-Master, all in gold, fitted with a genuine movement, and with a potential value of 35,000 euros on the vintage watch market. The case was simply recast in gold and a copy of the original dial was made (see illustrations).

Apart from the fact that the reference and serial numbers had been rubbed out, another detail caught my attention: the screw-in caseback screwed in a quarter millimeter too far. I had never seen such a thing before and to my knowledge there was no reason why one caseback should screw in deeper than the others.

At this stage, I started looking into the matter and found a host of errors. I started a painstaking investigation, calliper in hand. The verdict: the case and the dial had never seen the inside of a Rolex factory. It was simply a fake watch, its case the work of a goldsmith. Any dealer could be taken in at first glance.

A Breguet branded watch serves as a final example. It's a magnificent vintage chronograph in gold sold for 37,000 euros. For legal reasons, it cannot be illustrated here. Seasoned experts collaborated to reach a conclusion following doubts about the authenticity of the watch. There was a slight difference between the color of the gold caseback and that of the case. The verdict was that the serial number was that of another watch of the brand that had been entered into the registers; a fabricated case, a poorly copied dial, original hands but not for that model, and a movement made up of three different calibers. In short, a superb watch at first sight, but worthy more of Dr Frankenstein than of a watch manufacturer. A complete loss for the purchaser: 37,000 euros, to which must be added the services of the Breguet factory, which makes such identifications possible by keeping records, like a few other brands, such as Patek Philippe. It should be noted that few manufacturers offer this kind of service.

To conclude, not everybody can be expected to be able to detect a fraud and to follow up every suspicion with a quick investigation. When buying a rare and expensive watch you have to consult an expert. If the expert can make an appraisal, he will charge you for it, but the risks are reduced.

You will find in these pages an example of a genuine fake of an extremely rare product of the Rolex company, the reference 6062. I chose this model to give you an idea of the lengths to which a forger is prepared to go in producing a counterfeit watch, however rare it is or difficult to make.

Opposite page: fake vintage Rolex Ref. 1675.

ROLEX
GMT-MASTER
SUPERLATIVE CHRONOMETER
OFFICIALLY CERTIFIED

Fake Rolex Triple Calendar Moonphase Ref. 6062

This rare Rolex Ref. 6062 calendar watch deserves our attention. In this instance, the counterfeiter has reproduced a watch with a price tag of more than 100,000 euros! Judging by the state of the watch, everything is fake. It is already of a certain age and the case and bezel are entirely made of gold.

Fake Rolex Triple Calendar Moonphase Ref. 6062

Fake Rolex Triple Calendar Moonphase Ref. 6062

The case has been entirely made by the counterfeiter (in real gold). The shape is different, the most obvious visible detail being the lugs [1] that show a lack of material. The crown is not the original and a contemporary crown has been added (note the difference in color).

Original Rolex Triple Calendar Moonphase Ref. 6062

Fake Rolex Triple Calendar Moonphase Ref. 6062

The calendar [1] and moonphase [2] pushers are different on the two watches. The counterfeiter merely adjusted the pushers from another watch.

Original Rolex Triple Calendar Moonphase Ref. 6062

Fake Rolex Triple Calendar Moonphase Ref. 6062

Original Rolex Triple Calendar Moonphase Ref. 6062

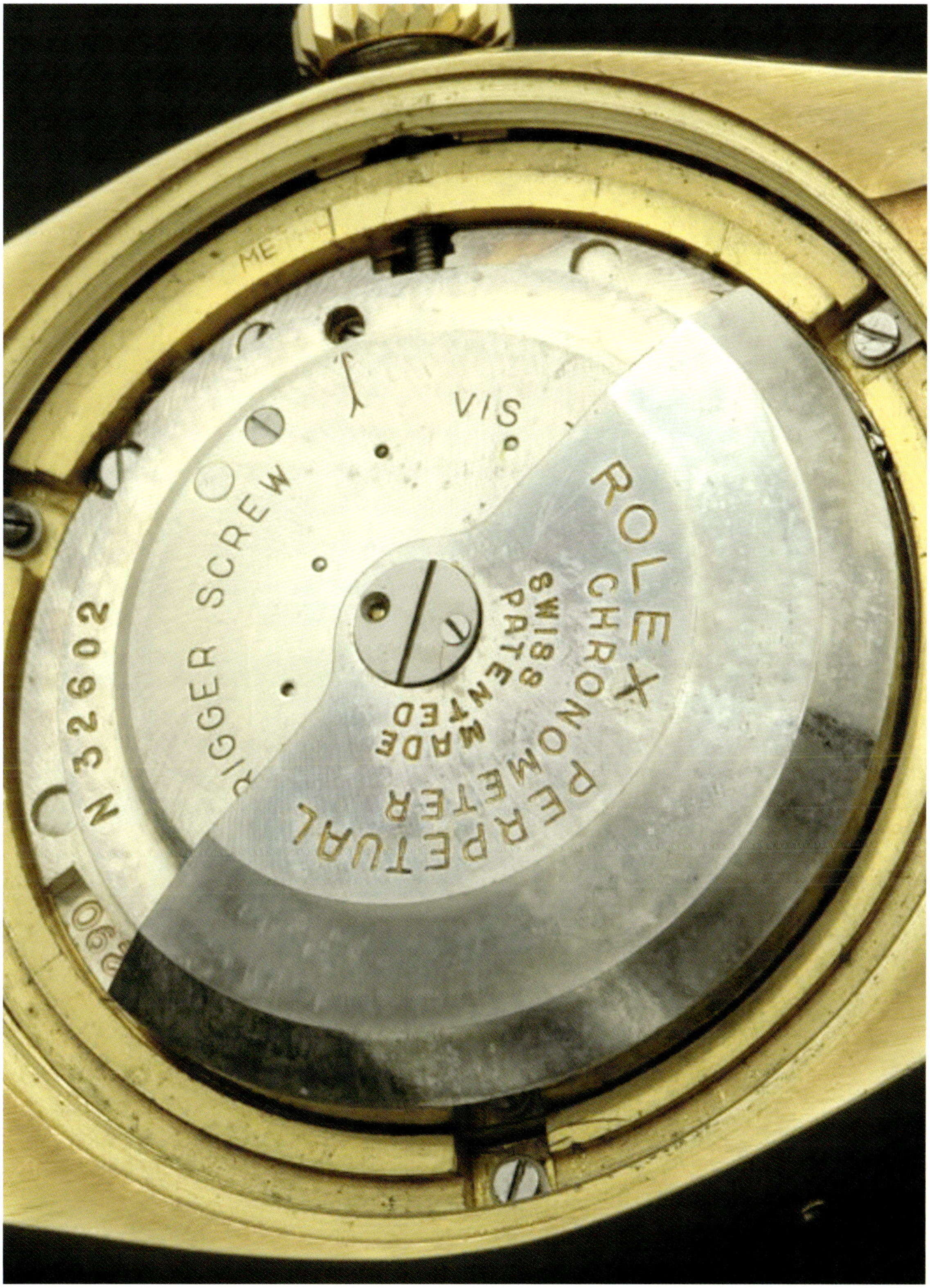

The mechanism is original. One might almost wonder whether this was the result of specially commissioned restoration work on a dilapidated watch performed by a restorer-watchmaker.

Fake Rolex Triple Calendar Moonphase Ref. 6062

A lot of differences appear when one taking a closer look at the details. A poorly placed fake crown [1] reveals the two holes designed to house the original crown. The seconds scale is also incorrect [2] and the moonphase features a different design [3].

Original Rolex Triple Calendar Moonphase Ref. 6062

The market for fake vintage parts

In the same way as they produce fake watches, some counterfeiters specialize in making fake components. The same situation already existed in the automobile parts sector.
These components include dials, cases, hands and other watch parts. In view of the increasing prices of some vintage models, it's not surprising to find counterfeiters specialized in making the fake parts that transform a series produced model into a rare object.
The difference in price between a vintage Submariner 5513 with a matt dial and the same model with a painted dial is several thousand euros. When dials are offered for less than 150 euros, it becomes a profitable operation for the less scrupulous vendor.
Apart from the trouble this creates for the unwary buyer, it needs to be pointed out that these practices have been going on for years with disastrous consequences.

Above: fake Rolex Submariner Ref. 6538 case bearing the reference between the lugs.

Opposite page: fake set of hands for the Rolex Explorer II Ref. 1655.

It is important for fans of vintage watches to remain vigilant, and it has become vital to become much more knowledgeable. If you want to protect yourself from being swindled, you now need to get acquainted with production dates, special editions, serial numbers, model changes and transitional models.

In fact, the problem is not new. How many among us will buy an 18th-century painting without making sure that it's genuine? And to do so you have to learn something about the history of art and painting techniques. Of course, this also applies to antique furniture and weapons or vintage cars. And the same goes for horology and timepieces.

Far from dampening your enthusiasm for vintage watchmaking, this chapter aims put you on your guard against this real problem. Increasing our knowledge of the subject enables us to enrich our watchmaking culture considerably, and for the more curious among us to definitively legitimize our interest in fine watchmaking.

It is now essential to be vigilant over each component of the watch, because the least doubt about any part it could result in a collectors' watch losing its value. When it comes to be serviced by the after-sales service, it could be destroyed as a fake watch.

Fake Rolex GMT-Master Ref. 6542 entirely machined by the counterfeiter.

Fake 1950s Rolex Ref. 5510 case.

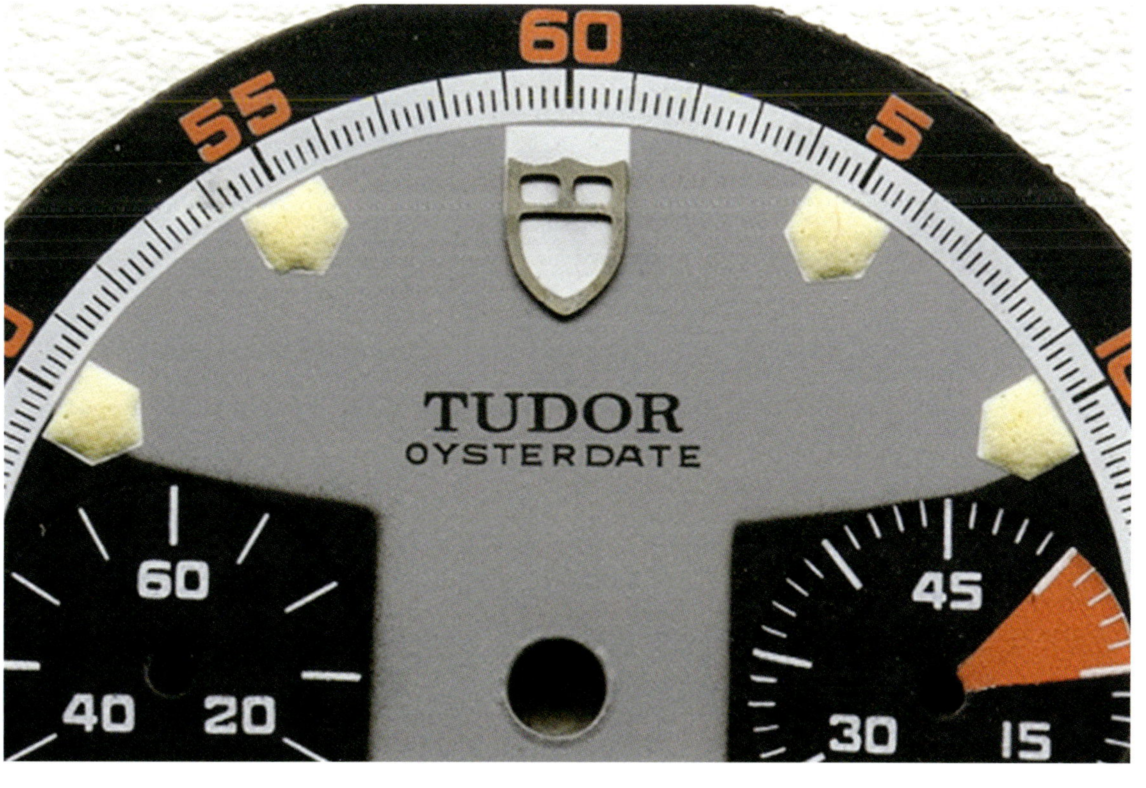

Fake vintage Tudor chronograph dial.

Fake vintage Rolex Daytona spare parts

Above: fake watch for the Rolex Daytona Ref. 6263 in 18K gold.

Opposite page: fake caseback of the Rolex Daytona Ref. 6263. Note the quality of the hallmark, even though this caseback is not made of gold.

Fake vintage Rolex Daytona spare parts

As if were not enough for counterfeiters to produce fake vintage watches by tinkering with parts taken from various watches, others are now specializing in producing fake spare parts. It turns out that they are inexpensive and well made, which means they are easily used by allcomers and lend themselves to fraud. Nonetheless, the quest for identification now implies becoming increasingly specialized and knowledgeable on the subject, a necessity for vintage watch lovers. The following pages feature a few comparative examples relating to a Rolex Daytona Ref. 6263.

Rolex Daytona Ref. 6263 case – Fake vs. Original

1. In the fake, the point is not centered on the 5 at 55 on the scale.
2. Letters shifted on the fake (the H crossbar centered on N in the original watch and shifted to the right in the fake).
3. The R of "Hour" shifted to the right in the fake. In the original, it is almost aligned with the P in "PER".
4. The point is located between the 1 and the 8 in the fake. On the original case, it is directly above the 8.

Fake

Rolex Daytona Ref. 6263 case – Fake vs. Original

5. The point is in the wrong place on the counterfeit case.

6 and 10. The graining is thicker and cruder on the fake.

7. The point is shifted to the left in the counterfeit.

8. The point is shifted to the left in the counterfeit. The scale is also in the wrong place (shifted to the left).

9. The point is slightly lower against the 5 in the counterfeit case.

Original

Rolex Daytona Ref. 6263 case – Fake vs. Original

1. The counterfeit case is made to house a caliber 7750 automatic movement or a manually wound movement made in China on a Lemania base. You can also find older Valjoux movements. These are today common for the great majority of counterfeit cases. However, be aware of the handymen who stock up with this kind of case and adapt them to accommodate a Rolex caliber 72 or 727 movement. Also consider that such watches have been on the market now for a few years. In this example, you can see that the cut of the case is not the same in the counterfeit.

Fake

2. The pusher is lower than in the original case. Also note that the notch of the chronograph pusher screw is cruder in the counterfeit case. Since the perforation is out of place, the pusher screw of the counterfeit case covers it entirely, whereas the upper section is fully visible in the original.
3. Finally, the perforation of the winder screw is too low in the counterfeit case.

Original

Rolex Daytona Ref. 6263 caseback – Fake vs. Original

This is an example of the screw caseback of the Rolex Daytona reference 6263. The most obvious difference is the grain of the gilding on the counterfeit, which is not real gold. The original is circular brushed. Furthermore, the diameter of the caseback of the counterfeit is smaller than that of the original. However, there are some fake pieces in which this detail is not obvious. Here, for example are the two covers shown for comparison.

Once again, the finish of the counterfeit watch is fairly unsophisticated.

Fake

Rolex Daytona Ref. 6263 caseback – Fake vs. Original

1. While the hallmarks certify the manufacturing quality and the grade of gold used in this watch are clearly visible, the counterfeit reveals a gross error. The hammerhead shows 12, while the original indicates 136, the licence number of the C.R. Spillmann SA company. Number 12 doesn't exist for a hammer without a handle, but it corresponds to the Genex SA company and should be represented by a key.
2. The hallmark, indicating that the gold used for the cover is of 18K quality, is not only unrefined on the fake, but you can also see that the badge around the face has the top left and right corners cut off, whereas the original badge is cut off on the left only and rounded on the right.

Original

Rolex Daytona Cosmograph Paul Newman dial – Fake vs. Original

The most alarming development in the production of counterfeits of this type of model is in the dials. Just two or three years ago errors in the typefaces were glaringly obvious.

Fake

Rolex Daytona Cosmograph Paul Newman dial – Fake vs. Original

Today, this difference is close to extinction. Here you can see the efforts counterfeiter has made to copy the typeface of the numerals on the subdials, including the thickness and the serifs on the ends of the numerals. No detail has been spared and this is just one of the products that can illustrate the progress their work has made.

But the counterfeiters still make overt mistakes so that a watch can be quickly and unequivocally categorized. In the example on the opposite page, the black part of the dial is not in relief as in the original watch (see the chapter above on the Rolex Cosmograph Paul Newman). The subdials are simply sunk into the metal. It is however important to note that there are certain "Non-step" (i.e. not in relief) Paul Newman dials on the market that are thought to have been originally made for Rolex and refused by the latter. This naturally adds to the confusion on an already extremely complicated market.

Original

Rolex Submariner Ref. 6538 "James Bond" dial – Fake vs. Original

1. The luminous point on the fake is a simple circle of white plastic. On the original it's a round piece of transparent Bakelite under which a layer of tritium is provided for night vision.
2. The winder marked "Brevet" on the fake is not the right shape and is too thick.
3. The teeth of the knurled grip of the bezel in the fake are too big and stick out too much. Furthermore, the original bezel is not in steel but in nickel-plated brass, which shows through as yellow when the plating is worn. The problem here is that this has to be verified and it is

Fake

obviously out of the question to scratch the owner’s bezel to check the type of metal. This model is really old and there’s very little chance of finding a new bezel out of stock without the slightest trace of wear. Nevertheless, when an original was sent for servicing, some bezels were replaced by steel service bezels.

4. The bars of the markers on the bezel of the fake are not of the right shape.
5. The typeface used on the bezel is wrong.

Original

Rolex Submariner Ref. 6538 "James Bond" dial – Fake vs. Original

1. The 5-minute graduations of the minutes scale are not of the right thickness in the counterfeit.
2. The points of the hour markers on the fake are nothing but ordinary layers of paint that have been grained and colored to create the effect of a patina. The original hour markers contained radium, a highly radioactive substance that quickly degraded the hour markers as can be seen on the original dial on the left.

Fake

Rolex Submariner Ref. 6538 "James Bond" dial – Fake vs. Original

3. The word Submariner is not at the right height. The typeface of the word "SUBMARINER" has mistakes that are visible on the enlargement, but it is getting ever closer to perfection.
4. The marker at 6 o'clock is of the wrong shape.

Original

Rolex GMT-Master Ref. 6542 case – Fake vs. Original

1. The Bakelite bezel is fake. There are a lot of so-called aftermarket or generic bezels of varying quality around. Many collectors and experts readily use a Geiger counter to measure the radioactivity of the bezel. The original bezels contain a large quantity of radium for night vision.
2. Fake contemporary winder (new from stock).

Fake

Rolex GMT-Master Ref. 6542 case – Fake vs. Original

3. The teeth around the bezel of the fake are not the right shape.
4. Furthermore, the original bezel is not in steel like the case. Wear reveals brass, a material that the bezel was made of before being nickel and chromium plated. Nevertheless, when an original was sent for servicing, some bezels were replaced by steel service bezels.
5. You can see differences in the quality of the reference number engraving under the cover.

Original

Rolex GMT-Master Ref. 6542 case (additional photos)

Below are a few additional photos of watchcases, with the fakes on the left and the originals on the right-hand page.

1. This watch isn't the easiest to categorize. The inscriptions on the caseback are exact, including the IV indicating the fourth quarter of 1957.
2. The serial number is correctly stamped at 12 o'clock on this reference (6542), but note that the original lacks the inscription "REGISTERED DESIGN".

1

2

3

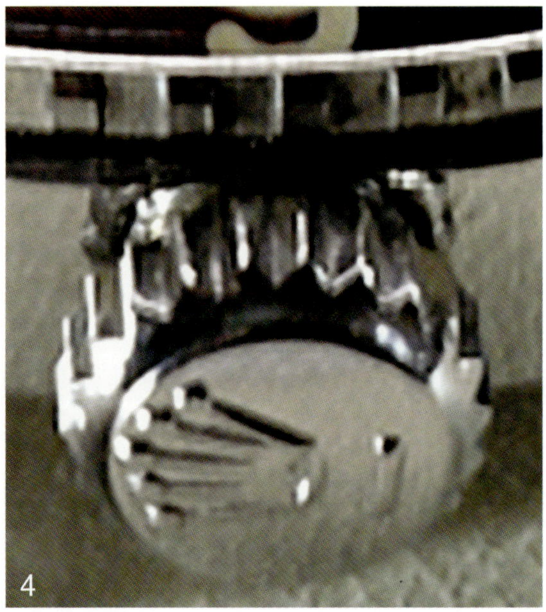
4

Fake

Rolex GMT-Master Ref. 6542 case (additional photos)

3. The teeth for turning the bezel to set a timezone are different from those on the original watch. The bezel on the fake is made of steel. It was made of brass for the original watch then nickel-plated. Age and wear allow the brass to appear. Nevertheless, when an original was sent for servicing, some bezels were replaced by steel service bezels.
4. The GMT-Master 6542 can be found on the market with a "Brevet" winder (cross beneath the crown), or with a dash beneath the crown. The winder with an underlined crown is simply a fake.

Original

Rolex GMT-Master Ref. 6542 dial – Fake vs. Original

1. The graduations of the scale are printed in relief on the fake with a thick, modern gold paint, which is not the case on the original.
2. The points of the hour markers on the fake are just plain coats of paint that have been grained and colored to create a patina. They are all of the same color and completely uniform.

Fake

Rolex GMT-Master Ref. 6542 dial – Fake vs. Original

3. The crown of the Rolex logo is not at the right height.

4. The anomalies in the typeface of "GMT-MASTER" are visible in the enlargement, but it is approaching perfection.

Original

22
CHRONOMETER
SWISS

When the counterfeiter goes into extreme detail

Let us now return to the same GMT-Master reference 6542 to examine an example of an artificially aged counterfeit. When a lens is no longer enough, a professional camera can be used to expose the most accomplished fake watches.

Opposite page: fake vintage Rolex GMT-Master Ref. 6542.

Artificial aging of the dial – Fake Rolex GMT-Master Ref. 6542

The result is astonishing when the dial is observed on a 1:1 scale. Whether under direct light or in sunlight, the "tropical" effect is clearly present. Under a magnifying glass, the effect is almost perfect. It took a high-quality macro photo to see that in actual fact brown/green paint has simply been applied to the lacquer of the fake dial, thereby rendering the deception clearly and definitively perceptible.

ROLEX
OYSTER-PERPETUAL
MT-MASTER
CIALLY CERTIFIED
RONOMETER

Use of other original metals – Fake Rolex GMT-Master 6542

For a long period, counterfeiters generally used wrong kinds of materials in producing fake watches. For example, counterfeit GMT bezels were often made of steel instead of nickel-plated brass. However, brass and nickel have indeed been used on the fake bezel shown here. It is therefore very difficult to identify this model given its close similarity with the original. The imitation game has even been taken so far as to pumice the metal so as to reveal glimpses of the yellow metal beneath the chrome plating.

Aged and accentuated flaws on the inlay – Fake Rolex GMT-Master 6542

Strangely enough, it seems that counterfeiters find it harder to imitate the font of an original part than to reproduce a complicated part. This inlay does not comply with the original font.

When purchasing this type of model, one unfortunately doesn't walk around with five or six original bezels. The best instrument for identifying the bezel is a pocket Geiger counter. In its early days, the bezels of the GMT-Master Ref. 6542 featured radium, a radioactive product enabling the bezel to remain readable in the dark. Radium will remain radioactive for around 2,500 years and the Geiger counter will start to click if in the presence of an original bezel. Meanwhile, tritium has a lifespan of around just 15 years. It is also a radioactive product and replaced radium on Rolex watches after 1958, for health and safety-related reasons.

Polished to create a worn effect – Fake Rolex GMT-Master Ref. 6542

Here's an example of a new case featuring simulated wear. It is not logical to have an over 60-year-old watch with a worn dial and crown, fitted into an entirely new case. In this instance, the counterfeit has therefore brushed the case so as to achieve an "overpolished" and worn effect.

ROLEX
PERPETUAL DATE
SUBMARINER
1000ft = 300m
SUPERLATIVE CHRONOMETER
OFFICIALLY CERTIFIED
SWISS MADE

fake!

Learning to identify a "quality" fake

If identifying a fake were as easy as is made out on the different websites and forums, it goes without saying that any one of us can become an expert in an afternoon. That is not to say that each one of us could not become an expert or that the subject is too complex, but one must not forget that the counterfeiters have made astonishing progress.
Having been quite impressed in 2015 by the build quality of a fake Hublot tourbillon, I had the opportunity in 2017 of examining three new-generation counterfeits that could make life difficult for quite a few unwary watch enthusiasts. These three models are presented in detail in the following pages.
The purpose of this chapter is to present additional features to check so that we can become sharper in our observation of the differences between an original watch and a counterfeit.

Models

Opposite page: fake Rolex Submariner Ref. 116610LV.

ROLEX
PERPETUAL DATE
SUBMARINER
1000ft = 300m
SUPERLATIVE CHRONOMETER
OFFICIALLY CERTIFIED
SWISS MADE
28
10
20
30
40
50

Fake

Rolex Submariner Ref. 116610LV
Fake vs. Original

Original

Fake Rolex Submariner Ref. 116610LV

At first sight, the differences between the two watches are relatively hard to spot. The fake watch above features a difference in the serif (thickness) of the characters appearing across the four graduations at the bottom of the dial, but that is not enough to enable full identification.

Original Rolex Submariner Ref. 116610LV

Fake Rolex Submariner Ref. 116610LV

When looking to identify the characteristics of a watch, it is important to focus on appraising its finishes. Authentic brand watches are well finished and manufacturers adhere to extremely precise technical specifications.

Original Rolex Submariner Ref. 116610LV

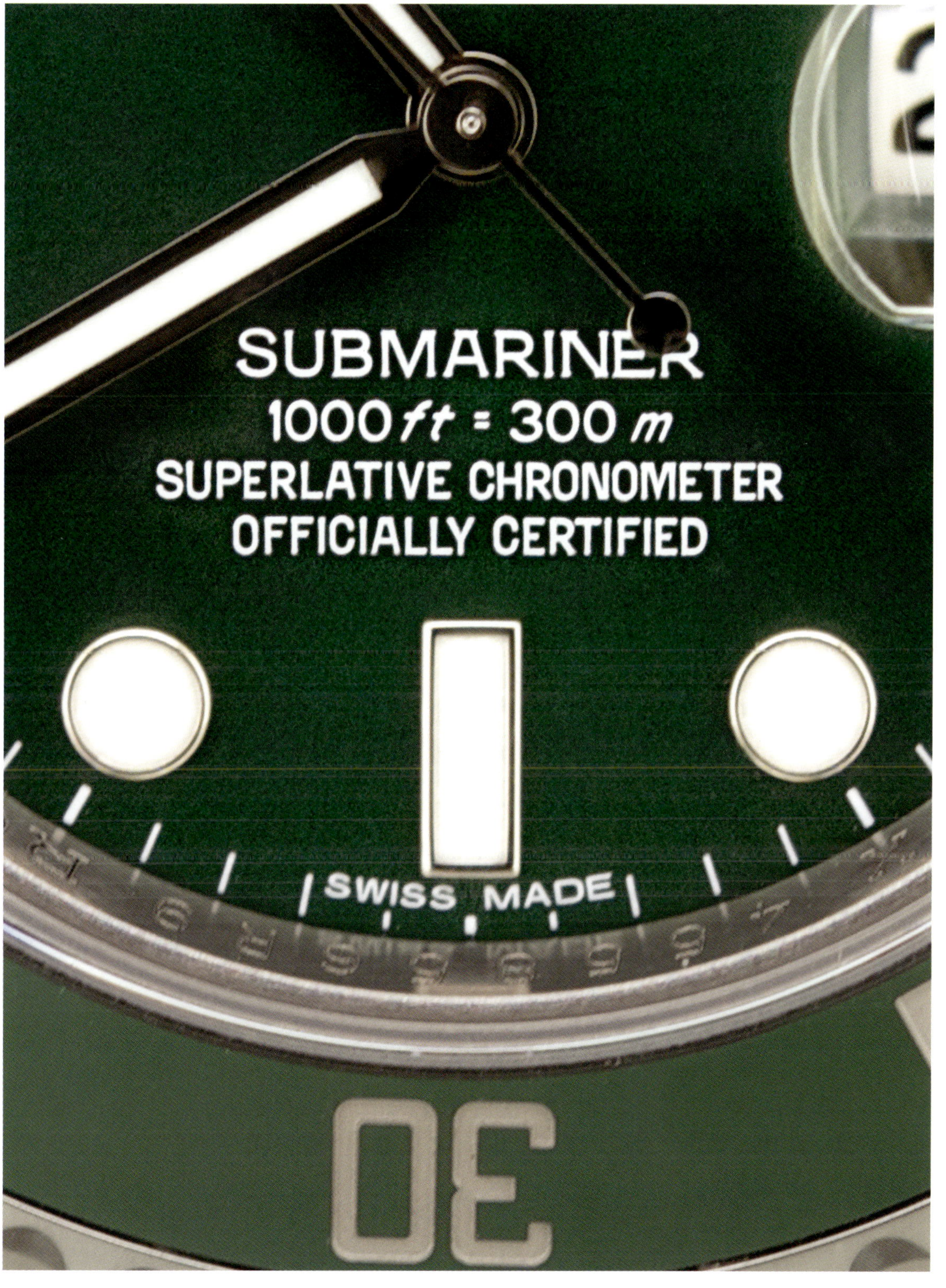

The term 'finishing' refers to the care for detail in the final touches lavished on each component before assembly. The fake watch appearing on the left-hand page shows poorly finished "ft" lettering as well as unequally thick hour-marker rings.

Fake Rolex Submariner Ref. 116610LV

There is an obvious difference in the shape of the hours hand, mainly visible in terms of the length and thickness of the lower bar.

Original Rolex Submariner Ref. 116610LV

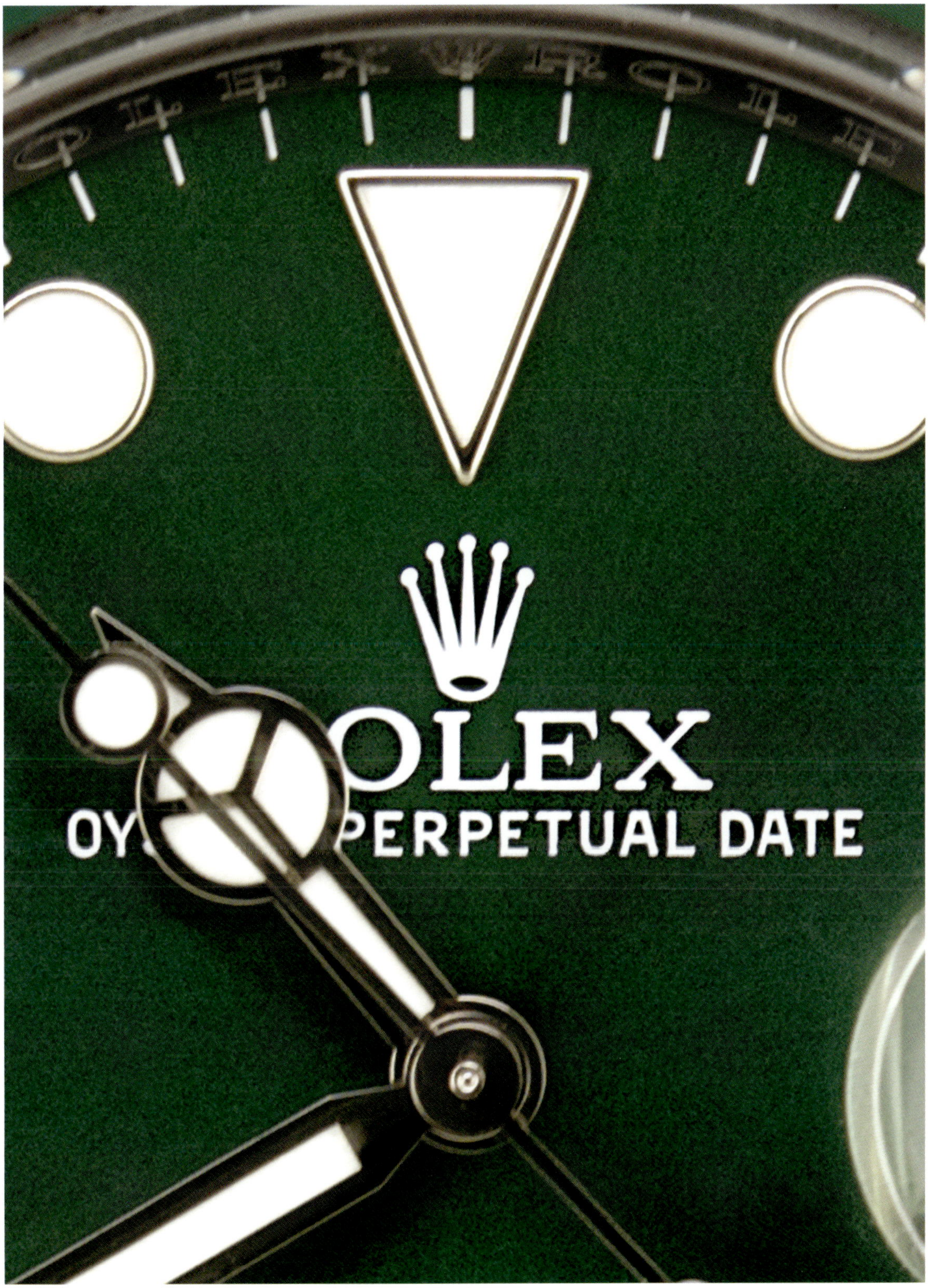

Fake Rolex Submariner Ref. 116610LV

When viewed under a magnifying glass, the marking on the date disk shows a visibly different level of quality. Take the number 2 of "29". The shape is different (with central rounding), as are the two ends. Bear in mind however that this may also vary from one production run to another.

Original Rolex Submariner Ref. 116610LV

Fake Rolex Submariner Ref. 116610LV

The bezel is relatively easy to identify.
The hollowed out portion of the numerals systematically leaves traces of machining, testifying to rough finishing.

Original Rolex Submariner Ref. 116610LV

Fake Rolex Submariner Ref. 116610LV

Another clearly visible detail: the marking between the watch lugs revealed as soon as the bracelet is removed.
The counterfeit features a laser engraving as opposed to the original bearing a real engraving.
See enlarged pictures on the following pages.

Original Rolex Submariner Ref. 116610LV

Fake Rolex Submariner Ref. 116610LV

Original Rolex Submariner Ref. 116610LV

Fake Rolex Submariner Ref. 116610LV

Three significant differences on this bezel detail:
[1] the quality of the engraving of the 12 o'clock triangle; [2] the luminescent capsule on the triangle;
[3] the notched rim (smaller notches).

Original Rolex Submariner Ref. 116610LV

Fake Rolex Submariner Ref. 116610LV

Difference in the quality of the markings:
the most visible detail is the quality of those on the crown, with the logo on the fake version appearing clearly very different from the original.

Original Rolex Submariner Ref. 116610LV

Fake Rolex Submariner Ref. 116610LV

Original Rolex Submariner Ref. 116610LV

Fake Rolex Submariner Ref. 116610LV

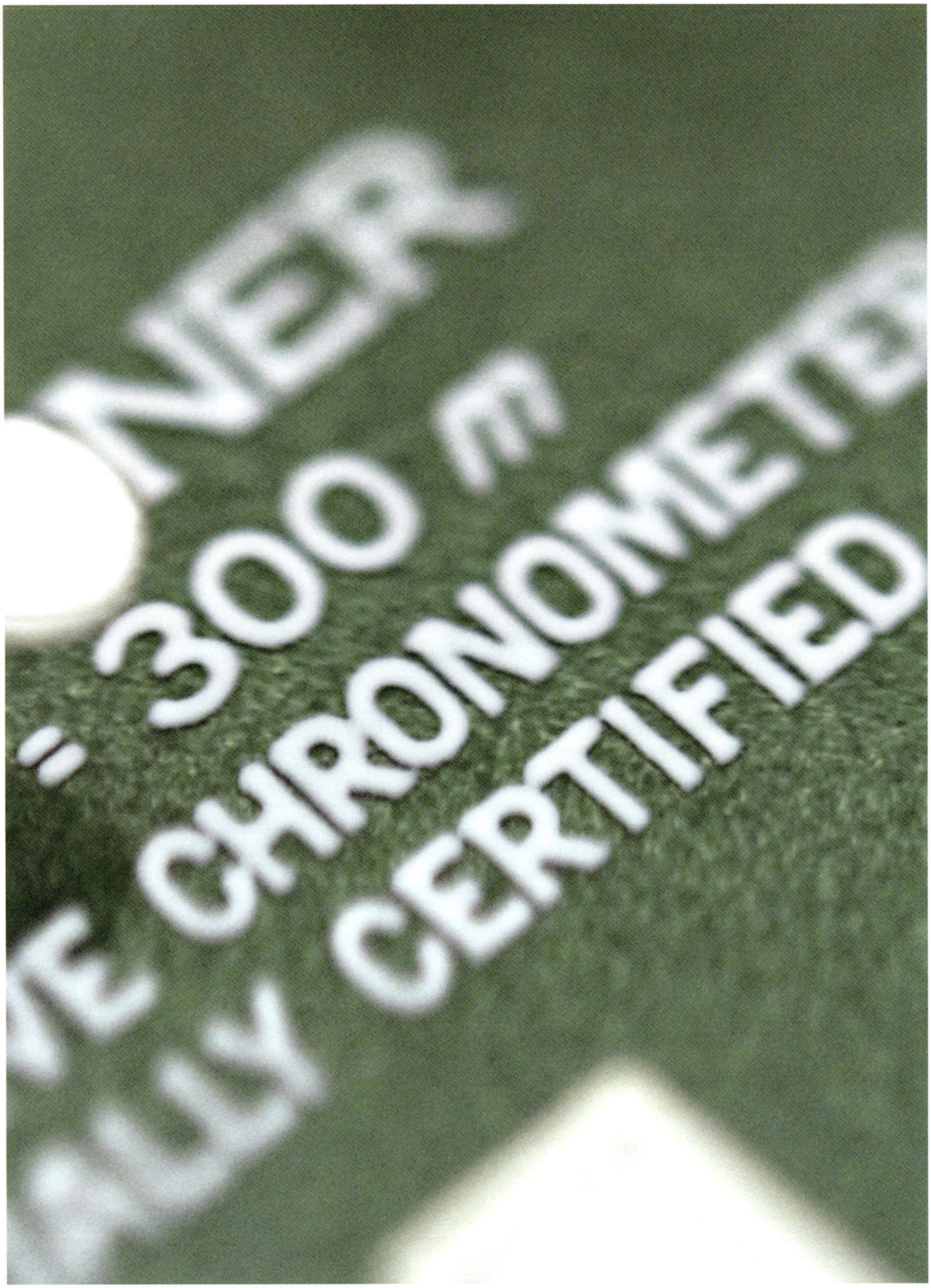

Font differences: the printing on the original Rolex
is incomparably more refined than on the fake.

Original Rolex Submariner Ref. 116610LV

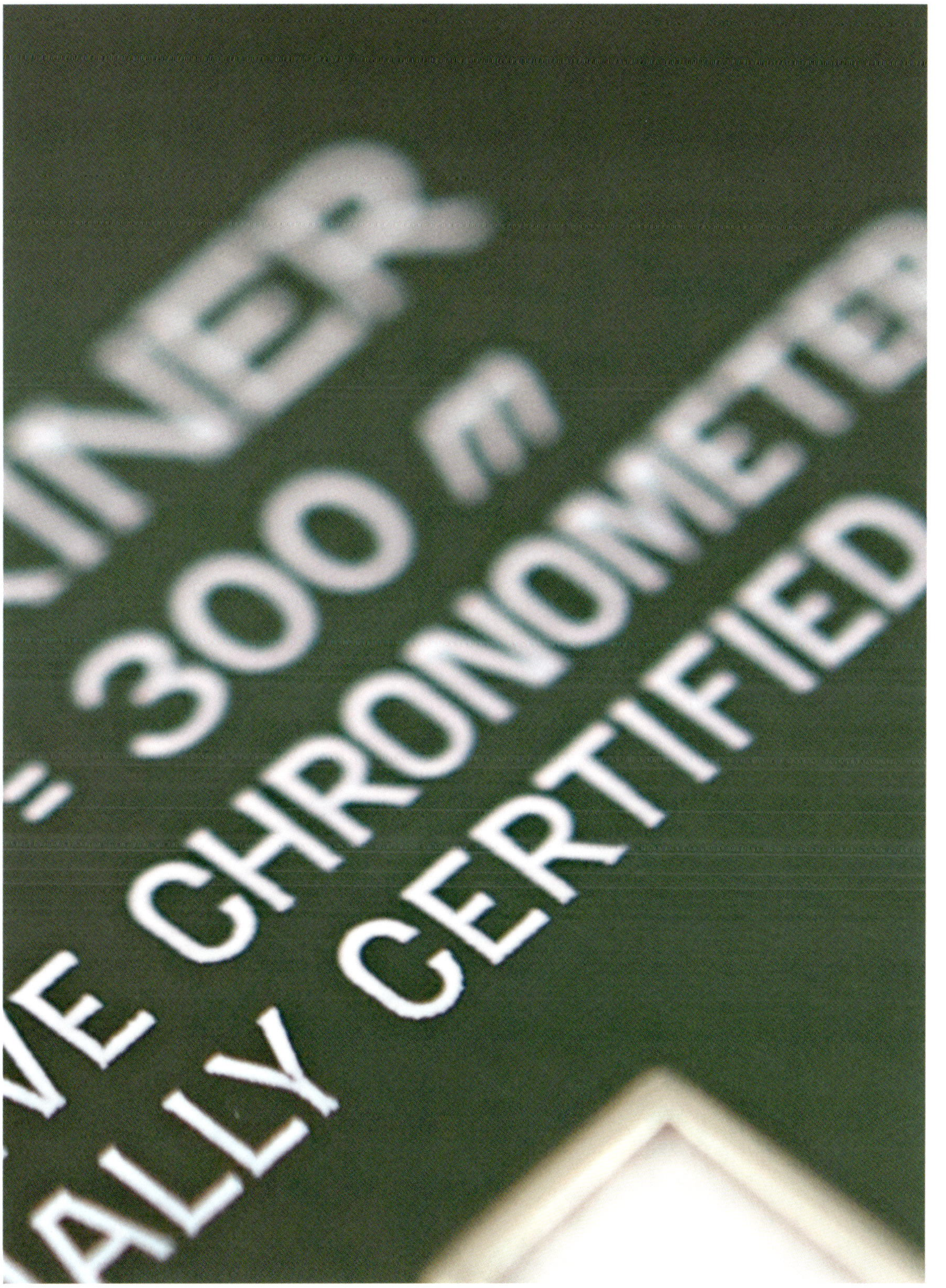

Fake Rolex Submariner Ref. 116610LV

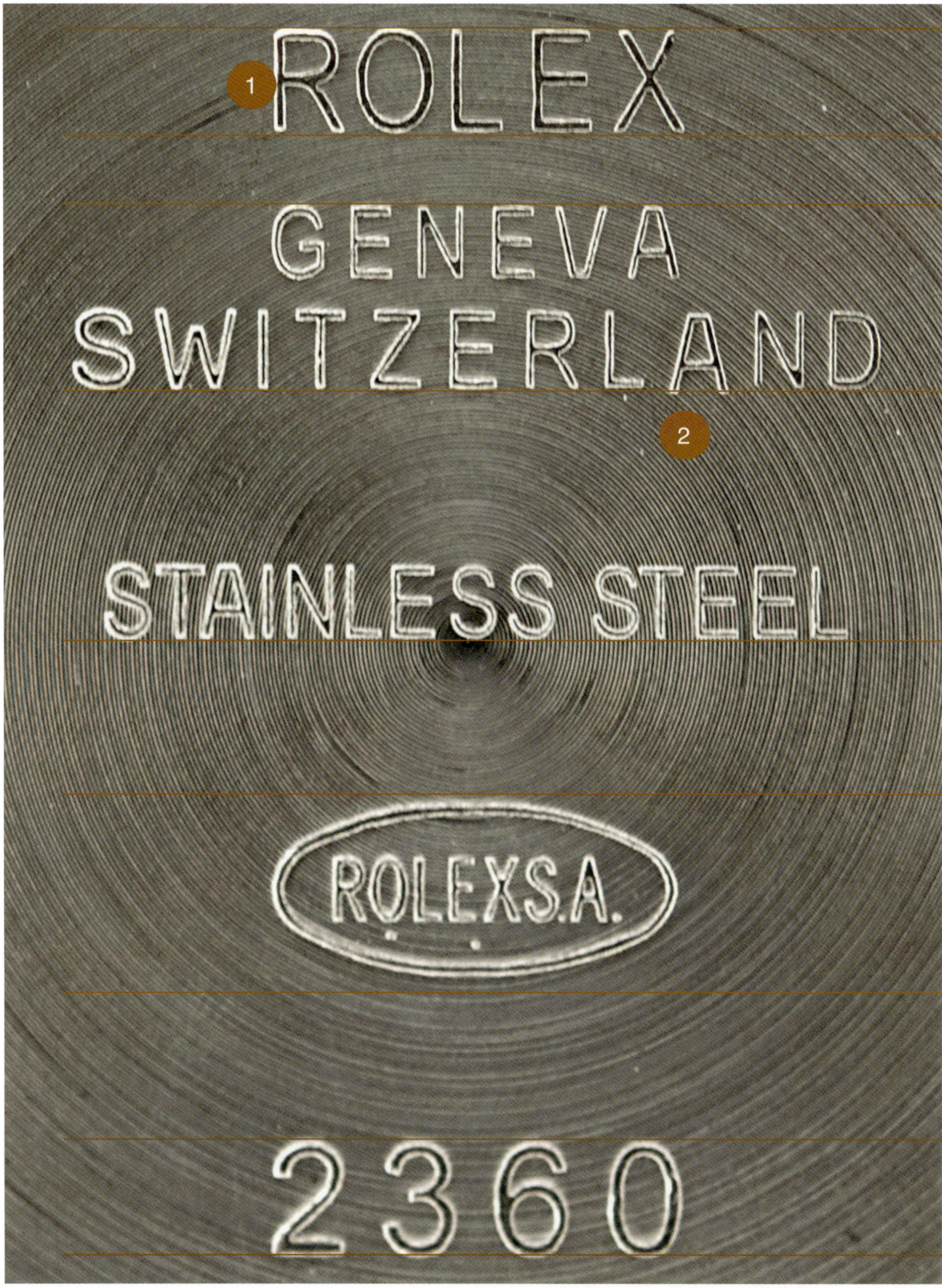

Example of an attempt at an exact imitation: there are few differences on the caseback, apart from the shape of the "R in Rolex [1] or the offset position of the "A" in Switzerland [2] compared with the alignment on the original.

Original Rolex Submariner Ref. 116610LV

Fake Rolex Submariner Ref. 116610LV

The movement is the element displaying the most striking differences in terms of finishes and production quality.
Take your time in observing these two photos and spotting a few differences in their quality of execution.
We will review the key differences between these two movements in the following pages.

Original Rolex Submariner Ref. 116610LV

Fake Rolex Submariner Ref. 116610LV

The most visible difference in terms of both aesthetics and finishing is found on the wheels serving to wind the watch automatically via the rotor, called reversing wheels (colored red on the brand's selfwinding movements). On this model, the counterfeiter merely placed a disk in the same color on a steel wheel.

Original Rolex Submariner Ref. 116610LV

Fake Rolex Submariner Ref. 116610LV

Another important difference lies in the balance-wheel system. Due to production costs, the counterfeit has opted for a balance with an adjustment index [1]. In terms of finishing, carefully checking the quality of the chamfers is an excellent means of analyzing the production quality of a watch [2]. Rolex has equipped its movements with a free-sprung balance fitted with inertia weights, which implies great precision and mastery of micromechanical engineering.

Original Rolex Submariner Ref. 116610LV

The original Rolex pictured above displays meticulous workmanship. Machining components in such an accurate manner is too costly for a counterfeiter. Rolex, which makes its own movements, will take as long as necessary to work on its parts so as to ensure perfect operation and an aesthetic quality worthy of its image, whereas the counterfeiter is interested only in achieving visual similarity and creating an illusion.

Fake Rolex Submariner Ref. 116610LV

One particularly effective exercise in getting used to comparing and observing is to take a quick look at two similar, yet non-identical parts. The ability to differentiate between similar an identical is the key to identification. I will point out a few and let you find the others.

Original Rolex Submariner Ref. 116610LV

Fake Rolex Submariner Ref. 116610LV

To get a more specific view of the field of watchmaking finishing and technical accomplishments, let's take a look at the different directions taken by the two stakeholders. The maker of fake watches merely needs to ensure the watch works. Rolex seeks to improve the quality of its products as well as their precision, their resistance, etc. In 2000, the company developed and patented an innovative hairspring made from an exclusive alloy composed of niobium, zirconium and oxygen: the PARACHROM® [1] hairspring. This component is entirely produced in-house. The aim was to produce a hairspring entirely unaffected by magnetic fields, while ensuring great stability in changing temperatures, and up to ten times more accurate than a traditional hairsping in case of shocks.

Original Rolex Submariner Ref. 116610LV

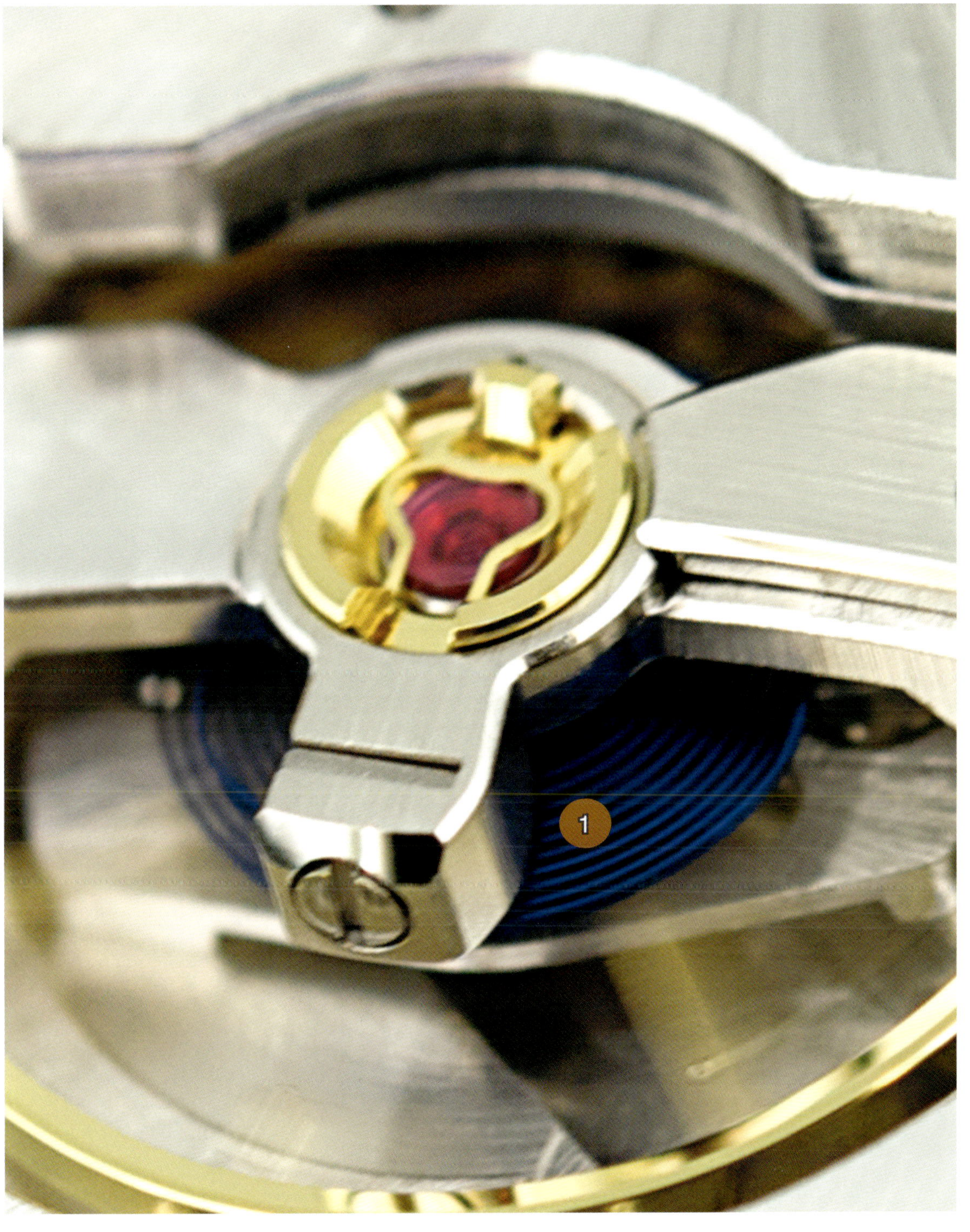

Five years later, Rolex introduced a new patented process that modifes the surface of the PARACHROM® hairspring in order to improve its long-term stability, a treatment that endowed it with its specific blue color.
As far as the counterfeiter is concerned, such a development would involve too great an investment and would increase the price to an extent that would drive it beyond a reasonable price range. At best, some counterfeiters may go so far as to anodize the hairspring with a blue coating similar to the original, but without ever achieving Rolex-level production quality. Production quality is definitely manufacturers' finest asset in the fight against counterfeiting.

Fake Rolex Submariner Ref. 116610LV

A good magnifying glass will enable you to see the difference in production quality on one of the most important parts of a watch mechanism: a screw. It must be perfectly machined if it is to hold two parts together. On a grand scale, imagine losing a cylinder while riding your motorbike due to screws of inferior quality. The same goes for horological micromechanics.

Original Rolex Submariner Ref. 116610LV

While Rolex demands perfect production quality when it comes to the screws equipping its movements, counterfeiters appear to underestimate the importance of this component. This is once again a question relating to production costs.

Fake Rolex Submariner Ref. 116610LV

Here we see that the fake screw is far better machined than on the previous page. That might lead to confusion if we were to forget that finishing quality remains our best asset in identifying an authentic watch. The screw on the original features a perfectly and accurately executed chamfer (a ridge with beveled edges) just as one would expect to see on a bridge or plate.

Original Rolex Submariner Ref. 116610LV

Fake Rolex Submariner Ref. 116610LV

Here we see the complete absence of a part normally fitted to hold the balance bridge in place [1].

Original Rolex Submariner Ref. 116610LV

Fake Rolex Submariner Ref. 116610LV

One of the reasons why it's important to hone one's capacity for detailed observation is that you often find yourself confronted with missing movement sections or components [1]. One should always keep in mind that there are plenty of errors and that they are merely hidden so that the fake movement 'resembles' the original as closely as possible - albeit only from an aesthetic rather than a technical standpoint.

Original Rolex Submariner Ref. 116610LV

Fake Rolex Submariner Ref. 116610LV

Bracelets are not spared from this plague. Whether sold with the watch or separately, fake bracelets are easier to make than a fake movement. The following pages will serve as a guide in your observations. Different font [1], laser marking on the fake [2]; engraving on the original, different thickness [3].

Original Rolex Submariner Ref. 116610LV

Fake Rolex Submariner Ref. 116610LV

Original Rolex Submariner Ref. 116610LV

Fake Rolex Submariner Ref. 116610LV

Different thickness [1], glued applied part on the fake [2].

Original Rolex Submariner Ref. 116610LV

Fake Rolex Submariner Ref. 116610LV

Different "Fliplock" bracelet locking system [1].

Original Rolex Submariner Ref. 116610LV

Fake Rolex Submariner Ref. 116610LV

The Rolex crown is ignored in terms of the shape of the original logo.

Original Rolex Submariner Ref. 116610LV

Fake Rolex Submariner Ref. 116610LV

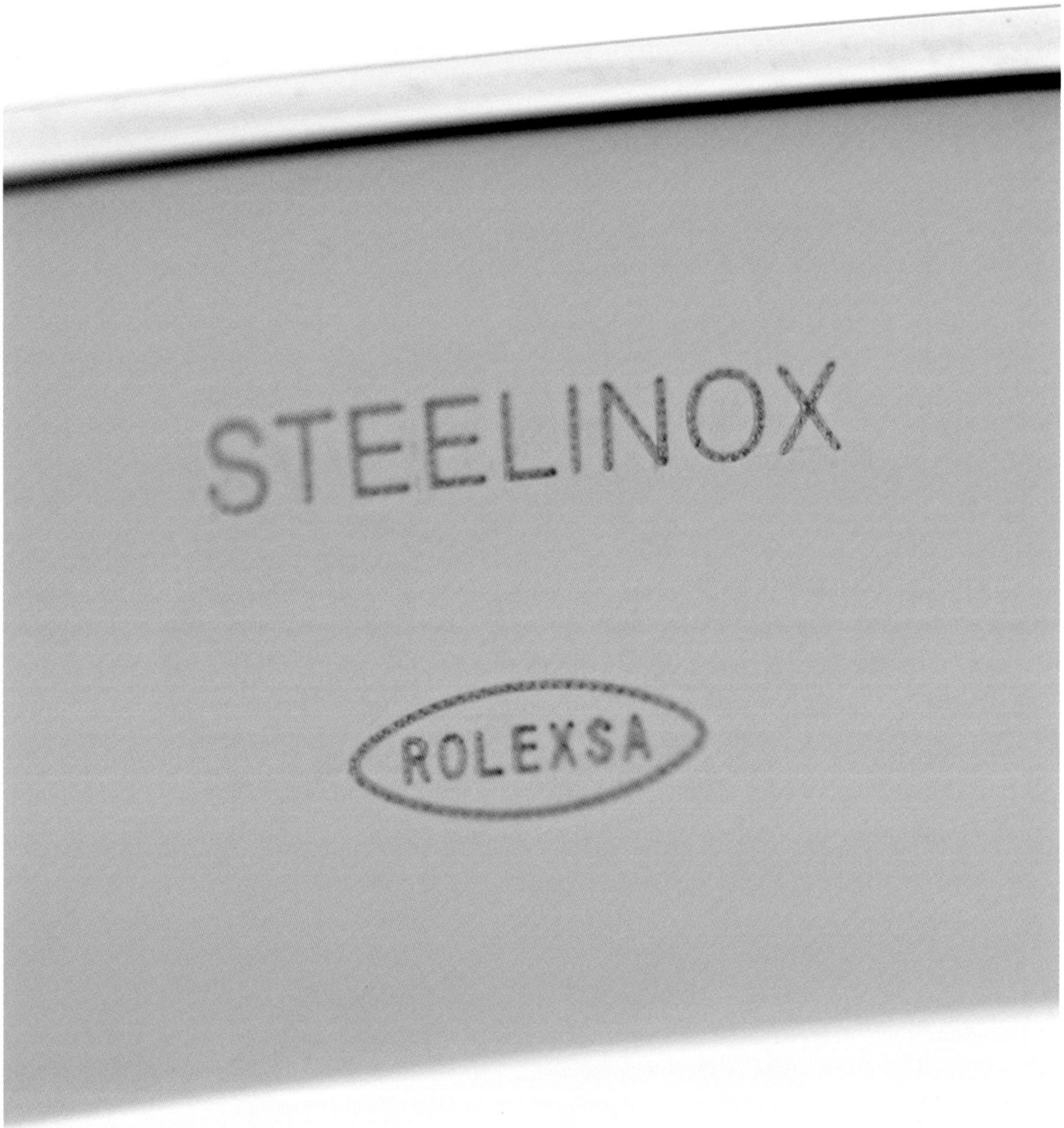

Just as with the markings of serial numbers and engravings on the bracelet lugs, the counterfeiter has opted to use inferior laser technology for the folding clasp of this model as compared with the original.

Original Rolex Submariner Ref. 116610LV

Fake Rolex Submariner Ref. 116610LV

The shape of this component, which serves to keep the watch bracelet firmly closed, is clearly different. While slightly beveled on the fake, it is perfectly rounded on the original. This detail will have an impact on the reliability of the bracelet fastening.

Original Rolex Submariner Ref. 116610LV

Fake Rolex Submariner Ref. 116610LV

Fast adjustment system for the size of the bracelet. Here too, the quality of finishing is different. The most blatantly obvious example lies in the absence of chamfer surrounding the pump-acccess opening [1]. Also note the difference in shape of the locking notches, with angles that are slightly rounded on the original Rolex and square-cut on the fake [2].

Original Rolex Submariner Ref. 116610LV

Fake

Audemars Piguet Royal Oak Ref. 15400
Fake vs. Original

Original

Fake Audemars Piguet Royal Oak Ref. 15400

The two watches look impressively alike, although the finishes are better on the original, particularly in the circular hollows of the manufacturer's famous checkered "tapisserie" motif. Simply rotating the watch is enough to note the rather dull gleam of the fake dial compared with the shimmer of the original.

Original Audemars Piguet Royal Oak Ref. 15400

Fake Audemars Piguet Royal Oak Ref. 15400

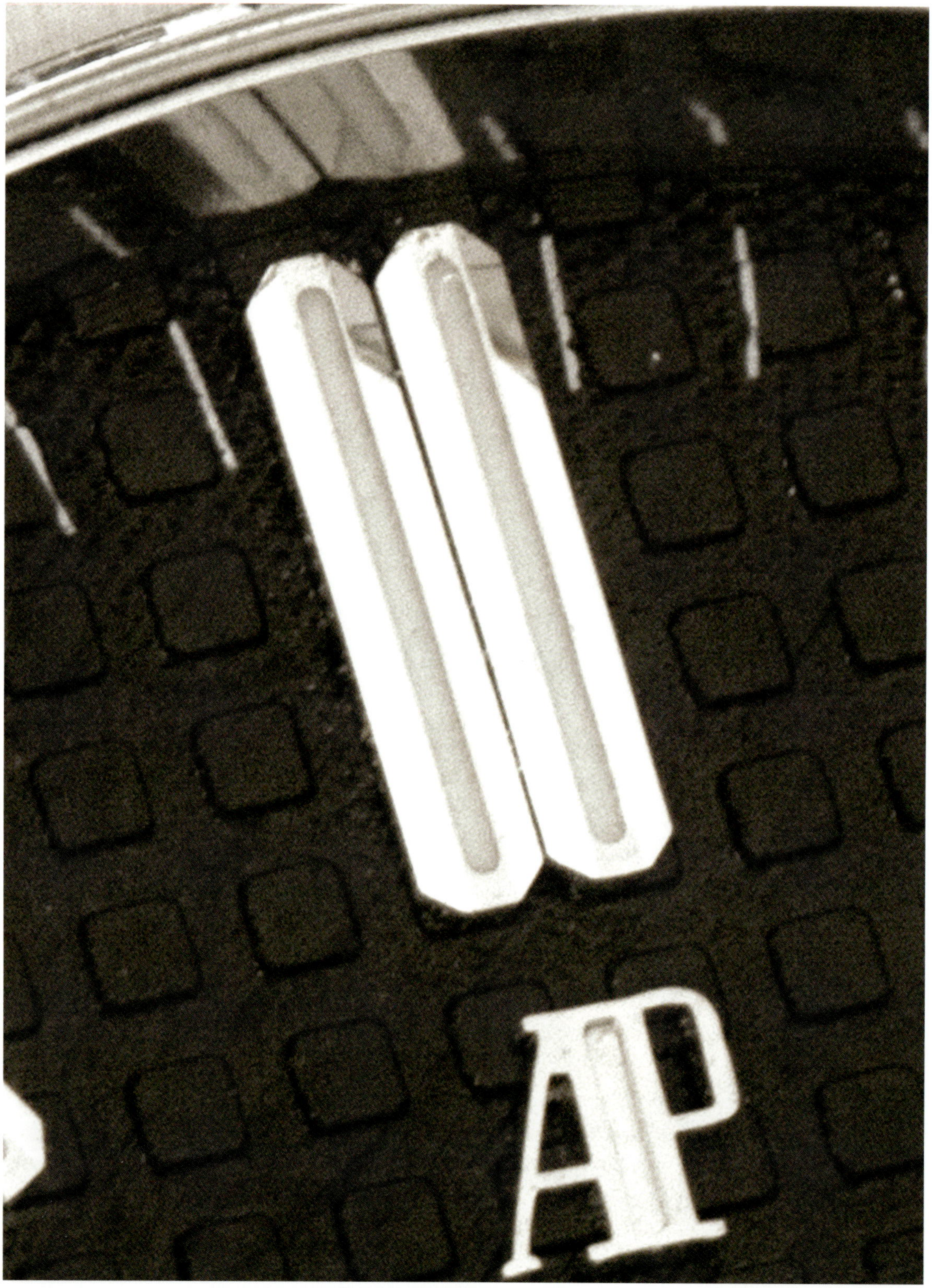

The refined nature of the painting on the original is clearly visible. The painting is also rather thick on the fake, thereby obliterating the effect of the "tapisserie" engraving on the original Audemars Piguet.

Original Audemars Piguet Royal Oak Ref. 15400

Fake Audemars Piguet Royal Oak Ref. 15400

The logo is not the right size and the finishing on the AP initials are not the same.
The font used beneath the logo on the fake is thinner.

Original Audemars Piguet Royal Oak Ref. 15400

Fake Audemars Piguet Royal Oak Ref. 15400

The date window is not the right size. The one on the fake is slightly more rectangular than the original. The font is bolder on the counterfeit, but on a 1:1 scale, these details are not so apparent and a x15 magnifying glass is required.

Original Audemars Piguet Royal Oak Ref. 15400

Fake Audemars Piguet Royal Oak Ref. 15400

The finishing quality of the hands is variable. Achieving perfectly beveled hands with optimal chamfers calls for a great detail of time and attention - two qualities the counterfeiter direly lacks.

Original Audemars Piguet Royal Oak Ref. 15400

Fake Audemars Piguet Royal Oak Ref. 15400

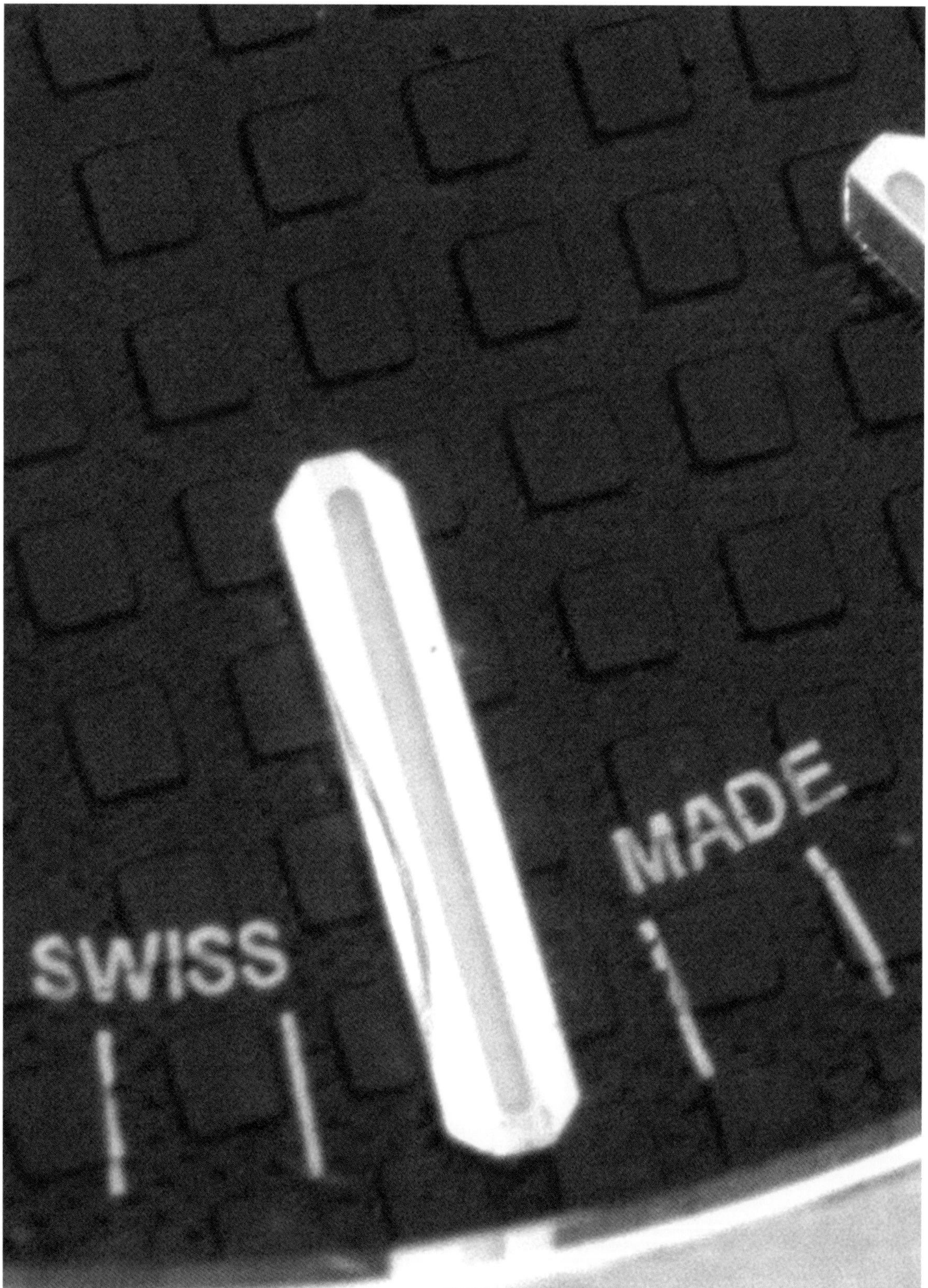

The difference in the print quality of the various markings on the watches is clearly apparent here.

Original Audemars Piguet Royal Oak Ref. 15400

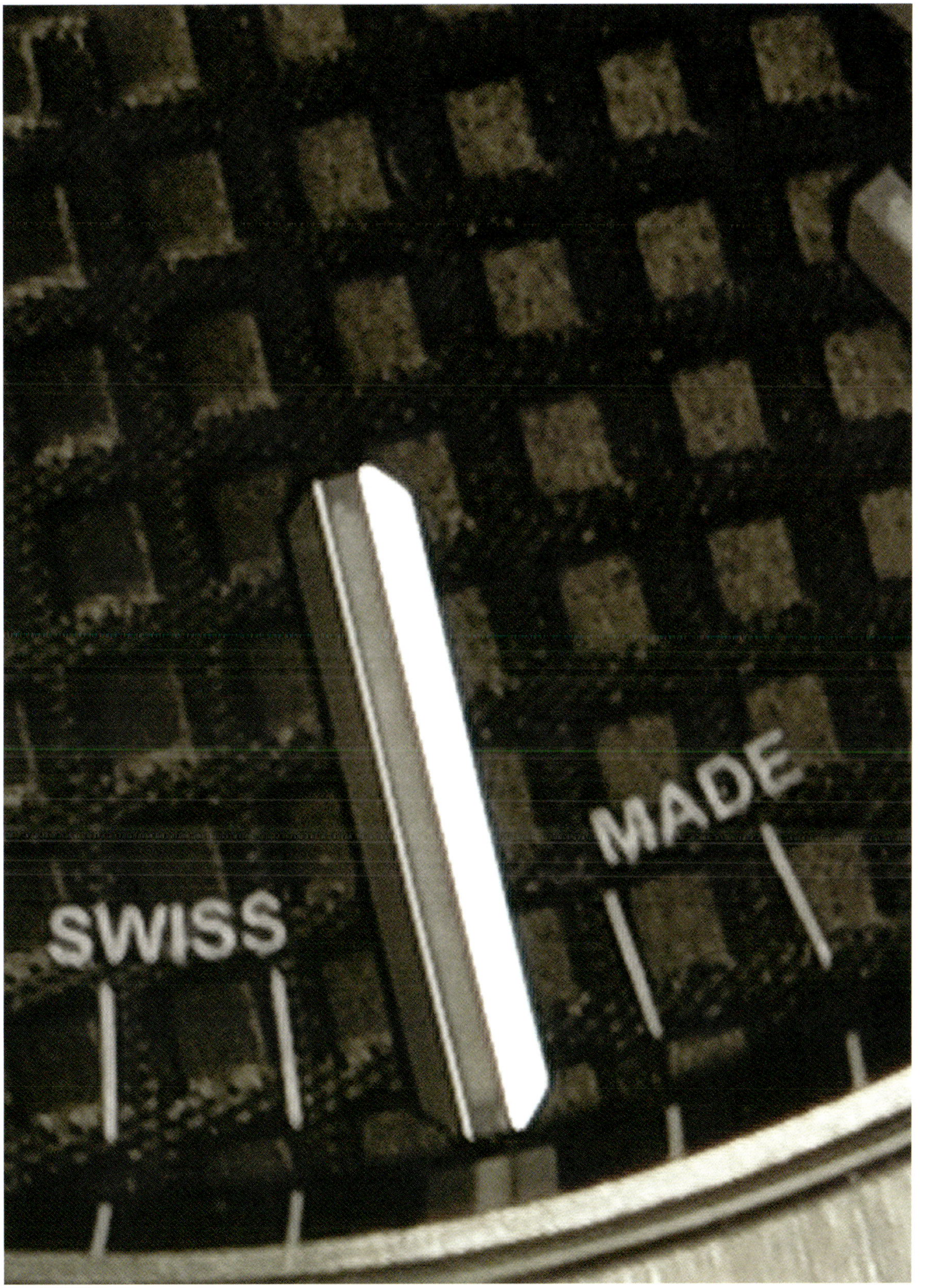

Fake Audemars Piguet Royal Oak Ref. 15400

One of the most dangerous pitfalls in counterfeiting lies in steel parts. Given that the same steel is used for both fake and original watches, these elements prove difficult to identify and these are not the parameters on which one should focus when conducting a comparison or an identification.

Original Audemars Piguet Royal Oak Ref. 15400

Fake Audemars Piguet Royal Oak Ref. 15400

For technical reasons, the counterfeiter will tend to let a mistake slip by. Here one notes a different thickness on the back of the watch.

Original Audemars Piguet Royal Oak Ref. 15400

Fake Audemars Piguet Royal Oak Ref. 15400

The steel oscillating weight of a fake watch is anodized to create gold-like effect (in this case pink, whereas the original is always in yellow gold). And the analysis does not stop there, so let's take an even closer look...

Original Audemars Piguet Royal Oak Ref. 15400

Fake Audemars Piguet Royal Oak Ref. 15400

Take the time to observe as many details as possible for yourself, in order to train your eye to spot the differences.

Original Audemars Piguet Royal Oak Ref. 15400

Fake Audemars Piguet Royal Oak Ref. 15400

The thickness of the engraving on the fake is extremely coarse. It depicts a horse which, despite looking fairly contented, is not much of a thoroughbred and not really worthy of appearing on the Piguet family crest.

Original Audemars Piguet Royal Oak Ref. 15400

Fake Audemars Piguet Royal Oak Ref. 15400

Viewed from a different angle...

Original Audemars Piguet Royal Oak Ref. 15400

Fake Audemars Piguet Royal Oak Ref. 15400

Original Audemars Piguet Royal Oak Ref. 15400

Fake Audemars Piguet Royal Oak Ref. 15400

These enlargements speak for themselves. Observing blatant insufficiencies in an imitation remains the best exercise for those wishing to identify a watch.

Original Audemars Piguet Royal Oak Ref. 15400

Fake Audemars Piguet Royal Oak Ref. 15400

If you look carefully at a balance bridge, as well as the balance wheel itself, you will note that not only are the finishes by no means the same, but also that the counterfeiters have felt free to "cut corners" on certain aspects - as we will see in more detail in the next double-spread.

Original Audemars Piguet Royal Oak Ref. 15400

Fake Audemars Piguet Royal Oak Ref. 15400

Audemars Piguet has equipped its movement with a free-sprung balance fitted with inertia weights [2 – right-hand page]. This implies great precision and mastery of micromechanical engineering – a method too complicated and expensive for the counterfeit, who in this instance has opted for a balance with an adjustment index [1 – on this page] for adjusting the balance-wheel.

Original Audemars Piguet Royal Oak Ref. 15400

Fake Audemars Piguet Royal Oak Ref. 15400

These two photos are a perfect example to illustrate the difference in finishing quality between a fake and the original.
The quality of the screws [1], the jewels [2], the chamfers [3] and the markings [4] is self-evident.
We also note the absence of certain functional parts on the fake [5].

Original Audemars Piguet Royal Oak Ref. 15400

Fake Audemars Piguet Royal Oak Ref. 15400

Another example of differences in quality (jeweling).

Original Audemars Piguet Royal Oak Ref. 15400

Fake Audemars Piguet Royal Oak Ref. 15400

A decisive factor in the identification of this fake watch: the visible mechanism of this watch is merely a decorative facade and not a full-fledged mechanism as on the original Royal Oak. While the ratchet wheel [1] and the pawl [2] were functional on the fake, they should be positioned facing the winding pull-out piece, which is not the case here. These are merely decorative parts intended to imitate the original caliber, just like many other superficial components. Dismantling the bridges and plates of the counterfeit would reveal a simple mechanical movement.

Original Audemars Piguet Royal Oak Ref. 15400

Fake Audemars Piguet Royal Oak Ref. 15400

Difference between the two logos on the crown. The one on the fake is much thicker.

Original Audemars Piguet Royal Oak Ref. 15400

Fake Audemars Piguet Royal Oak Ref. 15400

Let's examine the difference in quality between the link screws.

Original Audemars Piguet Royal Oak Ref. 15400

Fake Audemars Piguet Royal Oak Ref. 15400

Even though the bracelet seems to fit the fake watch, a closer look reveals that the finishes are very different, notably the size of the links [1] and the beveling [2].

Original Audemars Piguet Royal Oak Ref. 15400

Fake Audemars Piguet Royal Oak Ref. 15400

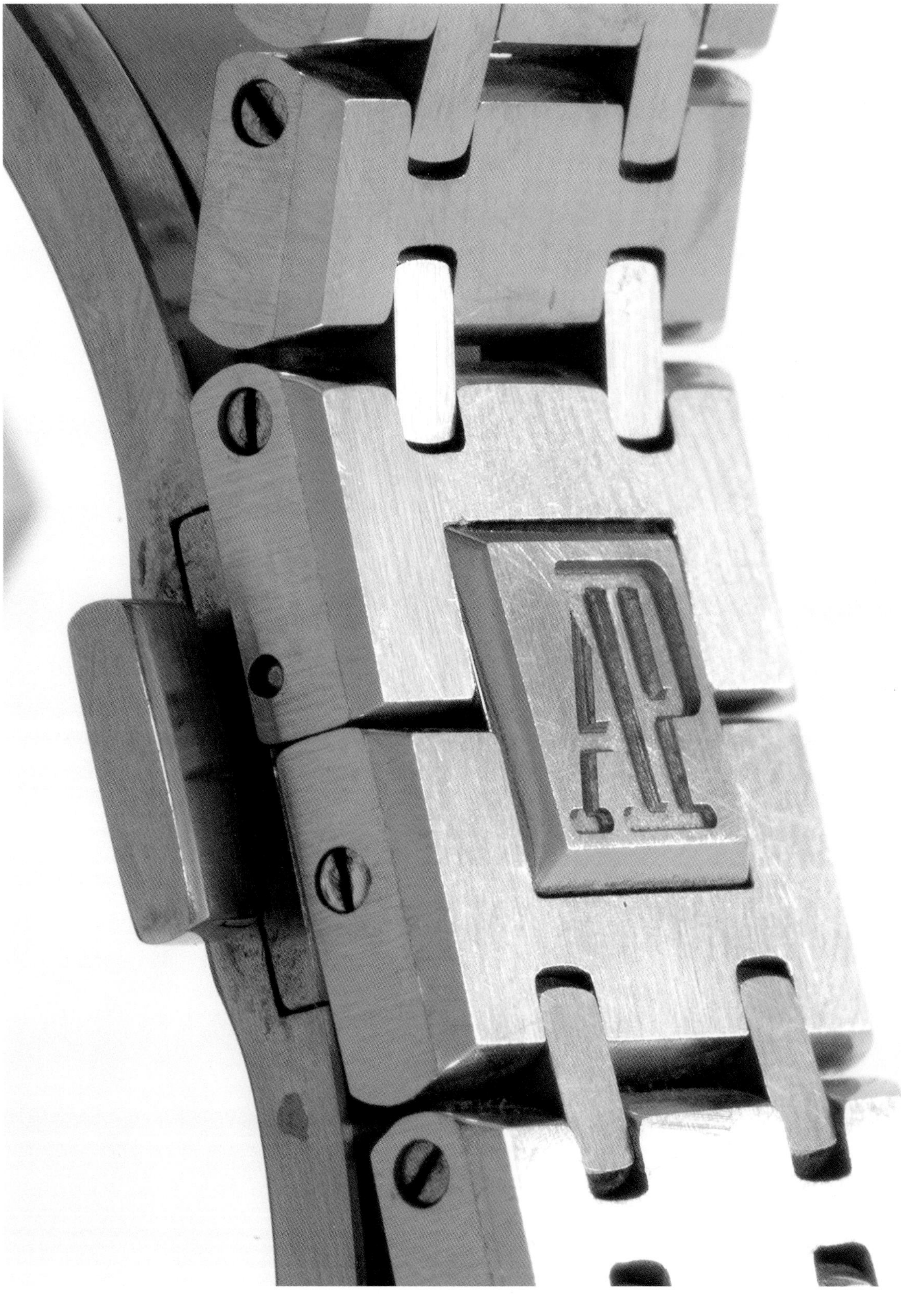

The AP logo plate is very crudely finished on the fake, as are the links.

Original Audemars Piguet Royal Oak Ref. 15400

Fake Audemars Piguet Royal Oak Ref. 15400

Original Audemars Piguet Royal Oak Ref. 15400

Fake Audemars Piguet Royal Oak Ref. 15400

While hard to make out on this photo, the locking system on the folding clasp of the fake [1] is not at all as reliable and perfectly adjusted as on the original watch. While the intention of the Manufacture Audemars Piguet is first and foremost to create a sturdy strap ensuring a firm hold on the wrist even in case of impact, the counterfeiter's intent is merely to imitate that part.

Original Audemars Piguet Royal Oak Ref. 15400

Fake Audemars Piguet Royal Oak Ref. 15400

Another illustration of the finishing and production quality of important components: the bracelet locking and closing system.

Original Audemars Piguet Royal Oak Ref. 15400

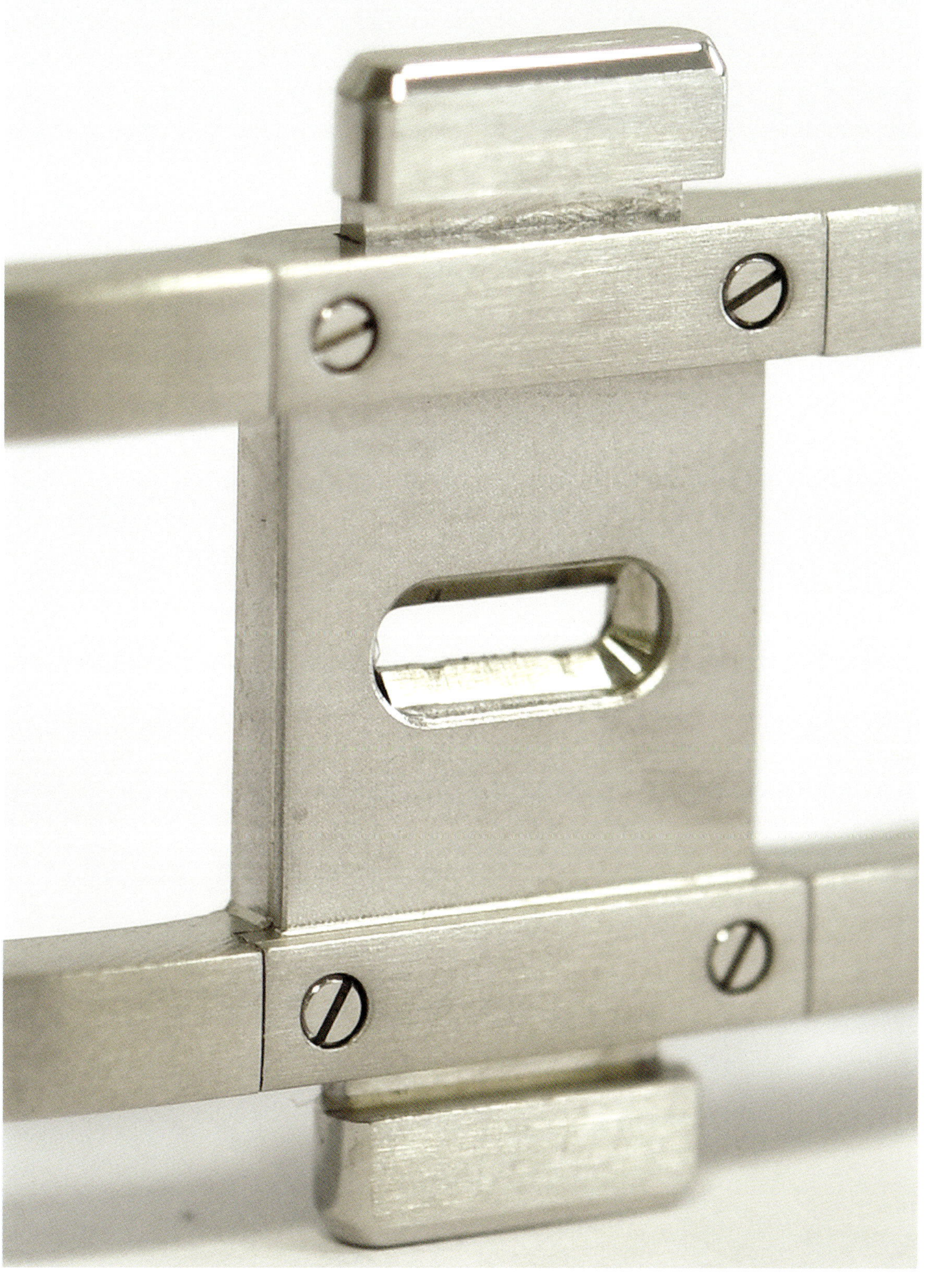

Fake

IWC Portuguese Automatic 7 days
Fake vs. Original

Original

Fake IWC Portuguese Automatic 7 Days

Different font thicknesses.
The fake dial is lacquered, while that of the original watch is slightly granulated.

Original IWC Portuguese Automatic 7 Days

Fake IWC Portuguese Automatic 7 Days

Difference in the thickness of the fonts: the paint used is entirely different, as illustrated in the close-up view on the following page. Granular paint on the original, versus raised varnish on the fake.

Original IWC Portuguese Automatic 7 Days

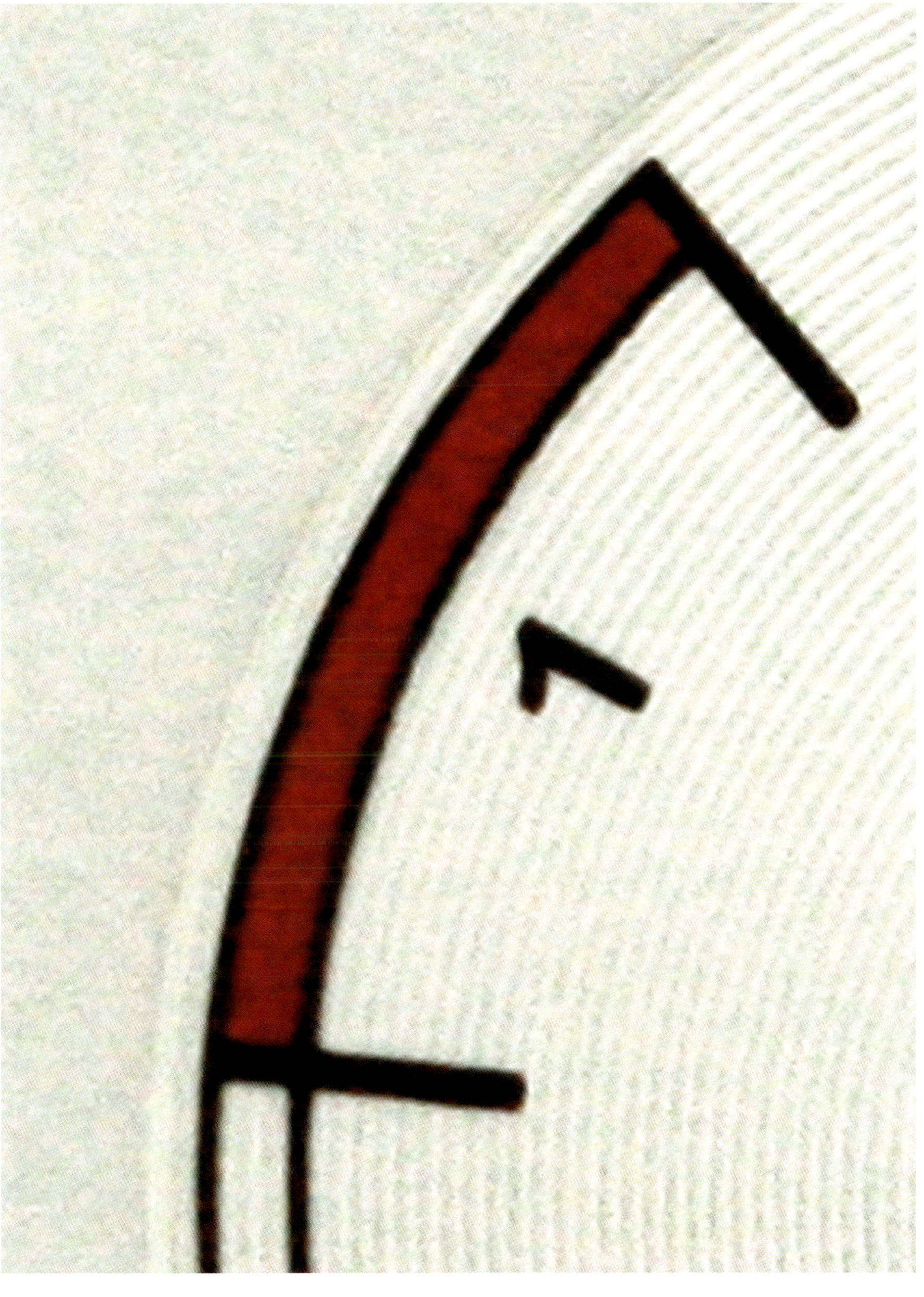

Fake IWC Portuguese Automatic 7 Days

Original IWC Portuguese Automatic 7 Days

Fake IWC Portuguese Automatic 7 Days

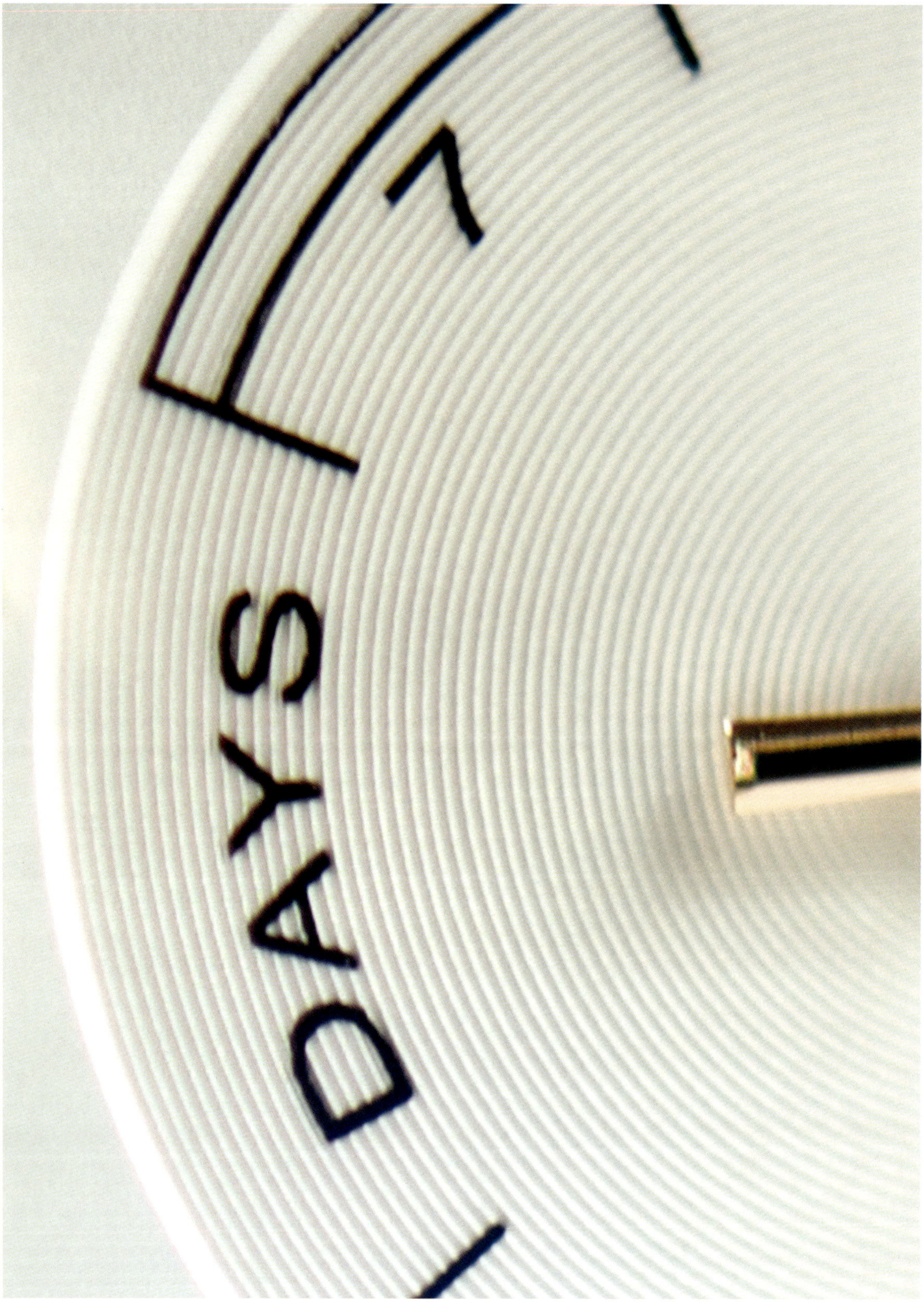

Difference between the thickness of the markings on the power-reserve display.

Original IWC Portuguese Automatic 7 Days

Fake IWC Portuguese Automatic 7 Days

Not only are the finishes on the hands of the fake of poor quality, but they are also not made of gold as on the original and are instead merely plated.

Original IWC Portuguese Automatic 7 Days

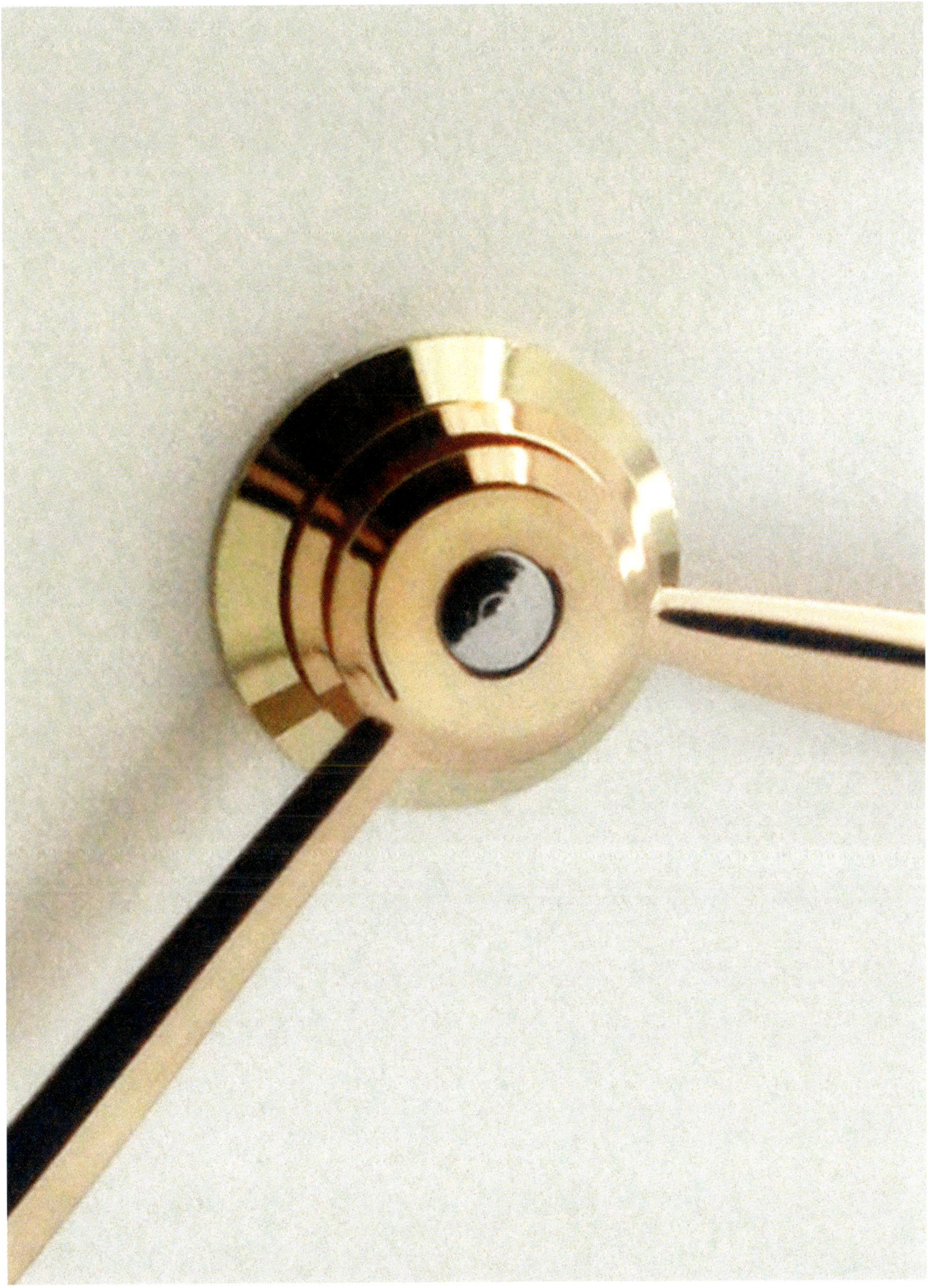

Fake IWC Portuguese Automatic 7 Days

The hand shaft clearly visible on the original shows a quality of finishing and a secure hold very different to those of the fake. IWC devotes a great deal of attention to beveling components, as can be seen here on the tip of the minutes hand (upper hand) [1]. On the fake, a simple straight cut is enough to create an illusion of authenticity.

Original IWC Portuguese Automatic 7 Days

Fake IWC Portuguese Automatic 7 Days

The gold hour markers on the original watch are riveted by several pins passing through the dial and secured by flattening against the back of the latter. On counterfeits they are often glued as shown here [1].

Original IWC Portuguese Automatic 7 Days

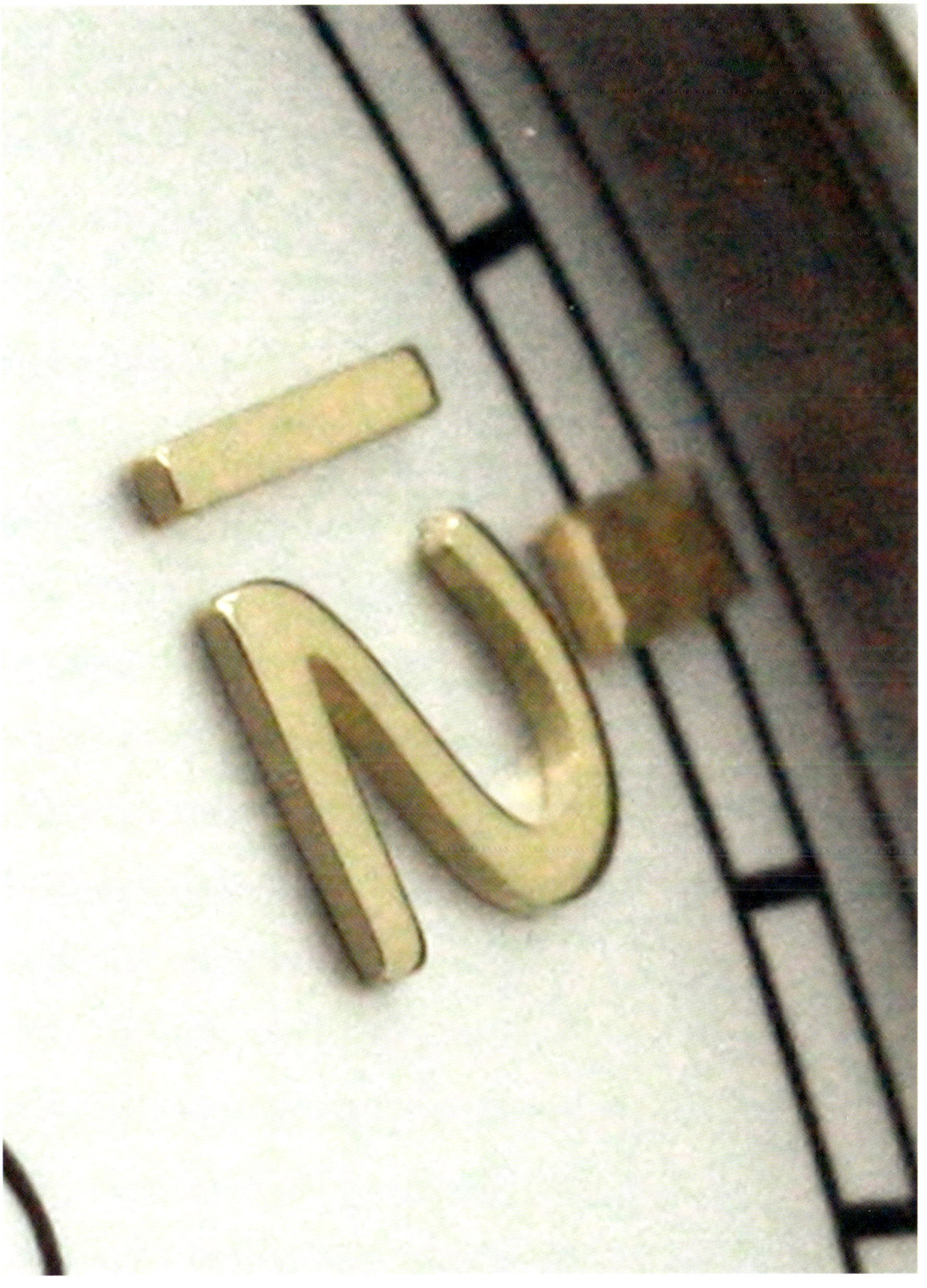

Fake IWC Portuguese Automatic 7 Days

Another example of poor finishing, visible under a magnifying glass [1].
The marker under the numerals is not the same shape as on the original watch.

Original IWC Portuguese Automatic 7 Days

Fake IWC Portuguese Automatic 7 Days

Simple non-polished cut angle on the fake [1] and polished angle on the original watch [2].

Original IWC Portuguese Automatic 7 Days

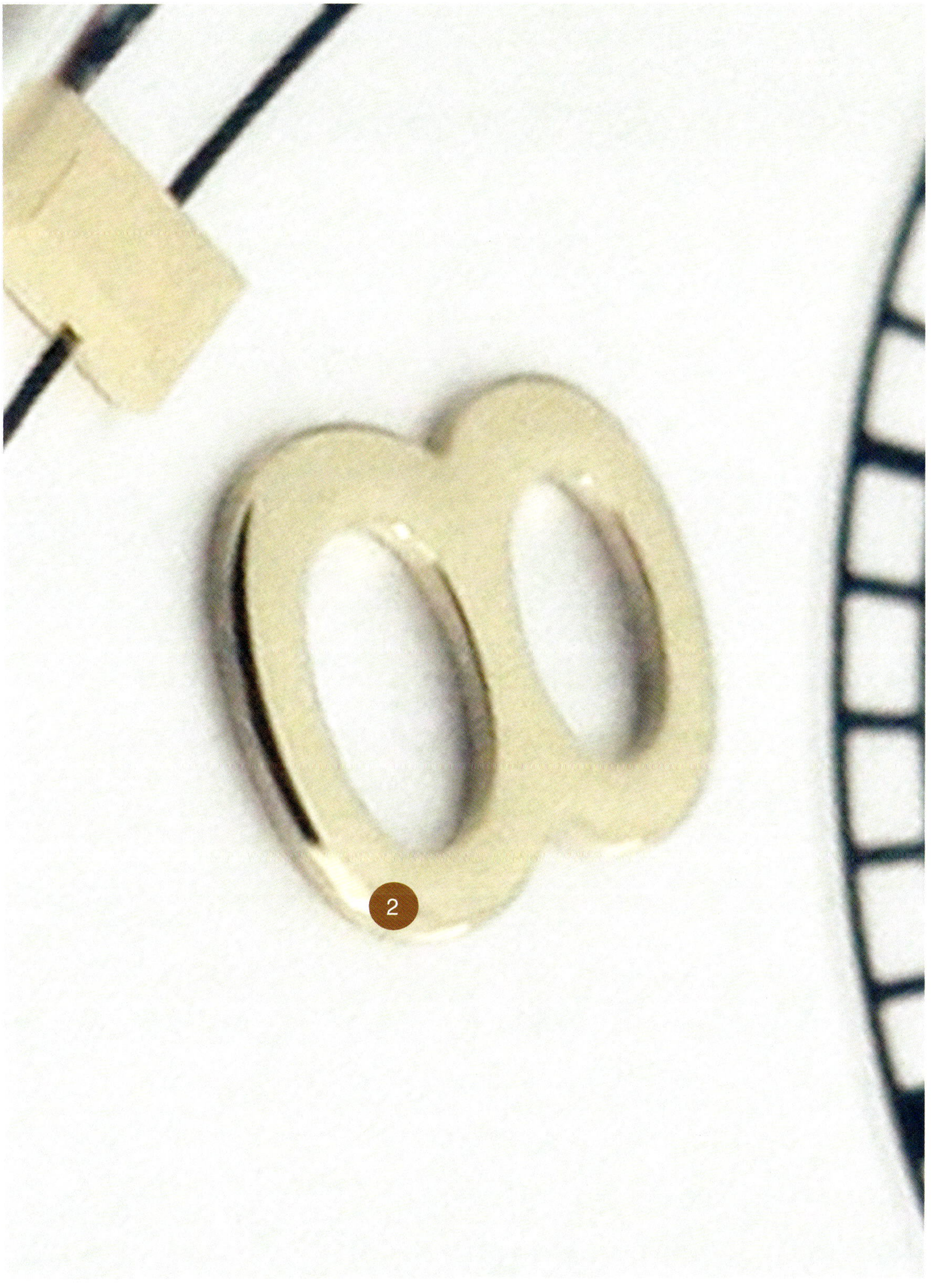

Fake IWC Portuguese Automatic 7 Days

The date aperture seems fairly well finished. It does however show glimpses of rough parts that IWC would have eliminated during the control phase of production [1]. Also note the different font used for the date.

Original IWC Portuguese Automatic 7 Days

Fake IWC Portuguese Automatic 7 Days

A fine example of the difference in terms of the shape of the part and the beveling.
The crown is shaped differently [1] and [2]. See the following page to observe the beveling differences.

Original IWC Portuguese Automatic 7 Days

Fake IWC Portuguese Automatic 7 Days

Original IWC Portuguese Automatic 7 Days

Fake IWC Portuguese Automatic 7 Days

Exotic leathers and skins: this fake features a poor-quality imitation leather.
There are two reasons for this: an alligator leather strap would cost the counterfeiter 50 times more than a fake steel bracelet.
Moreover, the exportation of exotic skins is severely restricted and stringently regulated.

Original IWC Portuguese Automatic 7 Days

Fake IWC Portuguese Automatic 7 Days

Let's take a look at the movement of these two watches. The brushed decoration is entirely different, as is the quality of the markings and above all of the parts. The most obvious example here is the winding wheel, which is in ceramic on the original watch and not on the fake [1].

Original IWC Portuguese Automatic 7 Days

Fake IWC Portuguese Automatic 7 Days

It is interesting to note that the counterfeiter even copies the movement references and goes so far as to have the movement components held by parts that are applied by gluing, rather than really serving to secure other elements (the screws do not hold the part in place) [1].

Original IWC Portuguese Automatic 7 Days

Fake IWC Portuguese Automatic 7 Days

Close-up view of the movement's winding wheel. On the left-hand side of the counterfeit sits a PVD-treated metal part.
The original watch features a precision ceramic wheel with a wheel-locking hook that is absent on the fake watch [1].
One may deduce that this part serves a purely decorative purpose on the counterfeit.
The wheel teeth are also differently shaped.

Original IWC Portuguese Automatic 7 Days

Fake IWC Portuguese Automatic 7 Days

It is very interesting to note that a "well-made" fake model always gives the impression of a perfect counterfeit. As part of the identification process, it is really necessary to ask oneself the following question: "Would the legitimate manufacturer have accepted this on one of its models?" Once again, as on the previous two fakes, one finds an adjustment index system [1] used by the counterfeiter instead of a free-sprung balance fitted with inertia weights [2]. From there, you have the right standpoint from which to make an objective comparison. Take a look at this part of the movement and judge for yourself.

Original IWC Portuguese Automatic 7 Days

As part of an identification process, it is really important to ask oneself the following question: "Would the legitimate manufacturer have accepted this on one of its models?" *[1] – [2] – [3]*

Original IWC Portuguese Automatic 7 Days

Fake IWC Portuguese Automatic 7 Days

In this instance, the wheels [1] and the quality of the jewels [2] are decisive indicators.
One also notes the inverted markings on the bridge which are not identical [3].

Original IWC Portuguese Automatic 7 Days

Fake IWC Portuguese Automatic 7 Days

The jeweling shows a clearly different level of quality.

Original IWC Portuguese Automatic 7 Days

Fake IWC Portuguese Automatic 7 Days

This example reveals some diverging production details. Considerable attention has been paid to the quality of the screws of the balance-staff on the automatic winding mechanism of the original watch. You won't find such are on a fake. [1].

Original IWC Portuguese Automatic 7 Days

Fake IWC Portuguese Automatic 7 Days

A typical example of rough finishes and the means employed by the counterfeiter [1], for whom finishes are not a priority, but merely a matter of visual effect. Conversely, a legitimate manufacturer will not allow such obvious flaws and will ensure that the model is perfectly finished.

Original IWC Portuguese Automatic 7 Days

Fake IWC Portuguese Automatic 7 Days

Original IWC Portuguese Automatic 7 Days

Fake IWC Portuguese Automatic 7 Days

Original IWC Portuguese Automatic 7 Days

Fake IWC Portuguese Automatic 7 Days

Easy-to-spot details: spelling mistakes [1], obvious error [2].

Original IWC Portuguese Automatic 7 Days

SUBMARINER
SUPERLATIVE CHRONOMETER
OFFICIALLY CERTIFIED
SWISS MADE

Protect yourself against counterfeit watches

by Alison Guéroux - Lawyer

This chapter offers a tool to help you anticipate or deal with any legal issue involving fake watches. However, jurisprudence teaches us that no case is to be taken at face value and that many factors can influence the outcome of the problem that confronts you. We recommend you read this chapter without waiting until you are victim of a scam, because it is better to anticipate potential litigation than have to face lengthy and expensive legal processes.

What is counterfeiting?

THE INFRINGEMENT OF INTELLECTUAL PROPERTY RIGHTS

"For sale: previously owned Rolex Submariner, 800 euros, very good condition with documents in original box." It looks like an attractive deal. You might be suspicious, but then there's still that small hope of being able to secure a fine watch at well below the market price. But you might be soon disappointed.

Counterfeiting has been a well-aired subject these last few years, justified by the fact that its presence on the market is ever increasing and that it not only affects the ordinary consumer but also collectors and dealers.

Counterfeiting is an infringement of intellectual property rights, in other words the fraudulent use of the exclusive rights on intellectual creations granted to the author of a literary or artistic work of the mind. The copy does not have to be exact because it need only be confined to one distinctive feature of the product to be counted as a forgery. Whether there are major or minor differences or whether or not the perpetrator is acting in good faith, a product can still be regarded as a counterfeit. The test is whether the copy bamboozles the consumer. The counterfeiting of manufactured goods is tantamount to deception because the counterfeiter tries to appropriate the creator's reputation thereby illegally benefiting from the investments made by the true owner of the intellectual property rights.

Within a few years counterfeiting has moved from a regional cottage industry to become a global industrial phenomenon. Counterfeiting has a major economic impact on society, accounting for 5% to 9% of world trade, or 200 to 300 billion euros. To increase profits, counterfeiters ignore the quality and safety standards of the products they copy. The increasing power of the Internet is the main driver of the increasing presence on the market of fake goods. It is indeed becoming ever easier to sell counterfeit products on "fake" Internet sites or with classified ads online, for in spite of various efforts to combat counterfeiters, many slip through the net. You therefore need to remain on your guard when acquiring a watch especially by keeping an eye open for some of the details that could help you identify a fake.

FAKE WATCHES IN THE COUNTERFEIT MARKET

It is clear that the global market in fakes has soared, rising from 100 million dollars to more than 250 million between 2000 and 2016. Thus trafficking in counterfeit goods accounts for 2% of global trade, according to the OECD, at the cost of almost 200,000 jobs, half of them in Europe. The annual production of fake Swiss watches is around 55 million units with an estimated turnover of a billion Swiss francs, or 5% of the sales of the 25 million genuine watches produced by the legitimate industry. Nearly two-thirds of watches on the market are therefore fakes, so there is every reason to be vigilant.

PARTICIPANTS AGAINST COUNTERFEITING

Faced with a problem of such magnitude, the Fondation de la Haute Horlogerie (FHH) has not remained idle. Out of its 500 members, 45 belong to the anti-counterfeiting group and 25 constitute the Internet cell. The FHH is active in several operations against counterfeiting, notably its annual video and poster competition on the theme "together against fakes" and its involvement in preventive campaigns.

While the FHH is a major participant in the combat against counterfeiting, others can also be mentioned. The 1901 TrueTime association, for example, is also active in the struggle. On June 14, 2014, the association launched a campaign with its sponsor, Stéphane Ciejka, entitled "Combat fakes and save our heritage." In addition to the *Revue des Montres*, several other international watch publications have become involved in the campaign, and up to the end of 2014 it was made known to watch enthusiasts around the world (United Kingdom, Middle East, Italy, Switzerland, United States, Germany and France).

Protect yourself from horological counterfeiting on the Internet

Given the quantity of fakes on the market and the rise in transactions between individuals on the Internet, it might seem difficult to avoid such a plague. But how can you tell it's a fake when you are buying online and you can't have the product in your hands? Here are some tips that could help you dodge the fakes.

To avoid the risk of being duped on the Internet, you should examine the features of the online outlet itself, such as the prices of the watches and the pictures.

DIFFERENT ONLINE OUTLETS

Sites dealing in fakes

Some websites are devoted entirely to the sale of fakes and advertise themselves as such. The message is therefore quite clear: the products are counterfeits, the customers know they are buying fakes, and you find such terms as "counterfeit, fake, copy, imitation" and especially "replica". Although such sites are clearly illegal, they give rise to the issues of their domicile and how to dismantle them. Many of them are registered under false identities, especially in countries in which no law exists to cover such contingencies.

The other problem is that if, despite these difficulties, the authorities manage to shut down these sites, they can immediately spring up elsewhere under a different domain name. The life of such sites is thus often ephemeral.
As it is clearly indicated on this kind of site that the watches are replicas, the customers know what they're getting. So you see arguments like: "Why pay for an original when you get the same thing for 20 times less?" or Swiss versus Chinese movements (the replicas are naturally not fitted with Swiss movements, nor with Swiss parts, but with counterfeit Swiss movements). If you buy watches for 150 or 200 dollars on such sites on the grounds of their deceptive appearance, the movements will be of poor quality, the watches will quickly cease functioning and you will be quick to arouse the suspicions of watch enthusiasts when you show them your latest acquisition.

Person-to-person sales sites

These cover all the Internet sites that connect individuals or dealers with one another with the view of buying and selling used goods. The best-known and the most popular for the trade in luxury watches include Vestiaire Collective, Leboncoin, eBay or PriceMinister (now Rakuten). In these sites, the goods are presented as genuine products. You might be led to think that you are buying from an individual who might be selling a watch to buy another or because he needs the money. Buyers feel more confident in these sites because of their reputation and the various protections (secure payment and shipping, assurances concerning fakes and the possibility of recourse to the vendor). The buyer is indeed much better protected in this kind of purchase than in the previous type of site because there are insurance systems for those who use such means of payment as PayPal.
Furthermore, in some of these sites the seller can be easily identified and so it is easier lodge a complaint. Each site might have options that give you varying amounts of protection, confidence in making payment, identifiable vendors or even a service to check that the product is authentic. All you have to do is to choose those that suit you best.
When it comes to second-hand watches for sale, you will notice a significant difference between the market prices of a new and a used watch — a difference that is easier to justify. Indeed, the new or second-hand products sold by individuals are generally cheaper than in the shop, and you can readily find authentic items of good quality at better prices. However, be careful because even for previously owned watches the prices might not be much lower than those asked by professional dealers. You can assess the reliability of vendors from their profiles or remarks made by other Internet users. You can also take a look at the other kinds of products that they are selling. Be suspicious if a vendor has several identical models for sale or if on the other hand if it is the first sale and his profile is still blank. Of course, there is always a first sale and the product might be genuine, but those who sell fakes often regularly create new profiles. You could also do an Internet search of a name or alias. Certain forums sometimes issue warnings of scams and mention the name or the alias of the vendor. Finally, you can make use of the vendor's photos; if they are insufficient or of poor quality, ask for others. Also note that the invoice and the packaging are positive factors, but they do not guarantee that the product is authentic, because if it is easy to fake an expensive wristwatch is it even easier to forge an invoice or make a fake box. There are many counterfeits on the sites for private sellers and buyers,

but some offer better protections than others and enable the vendor to be easily identified if there is a problem. The Vestiaire Collective or Chrono 24 Internet sites, for example, provide good security in the sale of horological products as well as a panel of international vendors.

SIGNS TO WATCH OUT FOR

How much it costs

The price is often a give-away. That is to say that a very attractive price is a sign that the watch could be fake, because unlike the major brands, counterfeiters spend nothing on research and development, marketing, quality control or even business premises. Above all they don't pay taxes or duties. Furthermore, a very large majority of counterfeiters employ, or rather exploit, black-market labor, very often children, in conditions that fail to meet human rights standards. They can thus achieve big profit margins even when they sell at slashed prices.

Taking it in hand

This is what we advise you to look out for if you have the chance of handing the watch:

Quality

The quality of counterfeit products is very often worse than that of the authentic item. If you're not sure or if you cannot claim to be an expert, you are strongly advised to ask the vendor to allow an appraisal by an expert or a retailer of the brand in question.

Packaging

The original box is often a good selling point, but once again it doesn't constitute a seal of authenticity, because if you can forge a watch, forging the packaging is easy. Therefore, it cannot be counted among the reliable aids to the identification of a watch.

Certificate of authenticity

Like the original box, certificates can also be forged. But the certificate can be used to detect a fake because in many brands its serial number should match that of the watch. You can then have it verified in the shop.

The penalties of buying a fake

What are the risks of buying a fake? The common scenario involves the purchase of a counterfeit product abroad or on the Internet followed by a customs inspection, but the penalties apply to anybody who has acquired a counterfeit. Thus, in addition to the seizure and destruction of the counterfeit object the person at the origin of the offense faces criminal sanctions.
Buying counterfeit products is punishable as a crime in France as in the European Union and many other countries. The government is actively engaged against the purchase of counterfeits and notifies consumers on the Internet of the features of counterfeit products:

- It usually involves organized crime and clandestine labor and illegal immigration networks;
- It prevents sustainable development by constraining local crafts to the benefit of imported copies;
- It is often linked with exploitation (absence of labor legislation, forced labor, child labor);

Counterfeit goods ignore all safety and quality standards. They endanger health (risks of al-

lergies or poisoning) and safety, especially fake medicines, drinks, perfumes and cosmetics, automobile parts and children's toys.
According to the law in France, regular customs checks take place at airports and ports receiving international arrivals as well as on packages sent from abroad. If you are in possession of counterfeits, you face two types of procedure (with the corresponding sanctions that could be cumulative in case of a large quantity of seized goods):

- Customs procedures: the counterfeit goods are seized and destroyed. The fines are proportional to the number of objects seized and calculated according to the value of the genuine articles. Such customs intervention is possible for the first counterfeit object, even though it is being worn. It is up to the owner to prove the legitimate origins of what he is carrying;

- Criminal procedure: in the same circumstances as above, the sentence is three years in jail and a fine of 300,000 euros, or about the same penalty as for theft. Counterfeiting is thus an offence that allows the suspect to be remanded in custody if necessary, quick trial procedures, and, in the event of sentencing, given a criminal record.

The purchase of a fake is not only the theft of an intellectual property right, it also means involvement in criminal networks, a contribution to the networks of illegal labor in conditions that violate human rights, the risk of buying a dangerous product, an economic nuisance and the risk of heavy criminal sanctions.

Consumer protection

The risks and penalties associated with the purchase of a fake are substantial, but if you are a buyer in good faith you naturally have no intention of buying fakes. However, in the case of a scam, you also have rights, especially the right to retaliate against the vendor. Consumer and civil law is designed to protect consumers. As a buyer you are protected in many ways by consumer law and civil liability in case of damage caused by third parties. Of course, the ideal approach is to take the maximum of precautions before you buy.

PROTECTIVE PURCHASING

You cannot take legal action against any person for counterfeiting unless you own the intellectual property rights. Nevertheless, you can always act against the seller to get your money back at least. If you realize that you have bought a fake:

- Consumer law should protect you: you don't even have to say it's a fake; if you send back the merchandise within seven days you get your money back, but you have to pay for the shipping.

- If you have used PayPal, which is a good method of payment against the risk of counterfeits: PayPal offers reimbursement if your purchase is a counterfeit. However, it is up to you start proceedings with PayPal and to declare a fake.

- Purchase on eBay: opening proceedings against the vendor. eBay acting as an intermediary undertakes to contact the vendor and to reimburse you. You only lose the shipping costs.

If you purchase in a shop, whether a retailer or a shop specialized in pre-owned watches, ask for an invoice and a certificate of origin. This will enable you to lay the responsibility on the shop for selling you a fake, even if they acted in good faith by believing they were selling you the genuine article.

CONSUMER PROTECTION LAW CONSUMERS' RIGHTS AND VENDORS' OBLIGATIONS

Consumer protection laws only apply to relations between professionals and consumers.
To be considered a professional, the vendor has to act within the framework of a professional activity, like selling products for example. Article L211-3 of the consumer code enables implementation of the right to guaranteed conformity in "the contractual relations between a seller acting in his professional or commercial capacity and a buyer acting in his capacity as a consumer." Not covered are the occasional private vendors or even transactions between professionals or between consumers. The consumer laws cannot be applied if you buy off a private person on a classified advertisements website or from an acquaintance.

The guarantee of conformity is applicable to the sale of luxury products from a professional to a consumer. The seller is bound to deliver a product that conforms to the contract and is liable for any lack of conformity existing at the time of delivery. Any lack of conformity appearing within six months of the sale is deemed to have existed at the time of delivery unless proved otherwise. Article L211-5 of the consumer code stipulates that to conform to the contract, "the product must be suitable for the purpose usually associated with such a product" or "have the features defined by mutual agreement between the parties or be suitable for any special requirement of the buyer which was made known to the seller and which the latter agreed to." If a defect is identified, the seller is obliged to repair the product or to replace it and even to reimburse the consumer. Action resulting from lack of conformity lapses two years after delivery of the product.

The right to withdraw
The consumer has the right to change his mind without giving any reason and without suffering any penalties. As a consumer, you have seven days after receiving the product to withdraw from the contract, return the product and get your money back. However, the consumer might be liable for the delivery costs.

Civil liability law
Unlike the consumer code which only applies to transactions between professional sellers and consumers, the civil code covers all contractual relationships including those between private people. You will thus come under the protection of civil liability law if for example you buy a watch from a private individual found on a site of personal classified advertisements.

Warranty against latent defects
According to articles 1641 and following of the Civil Code, the seller is bound to a warranty on account of the latent defects of the thing sold. The warranty comes into force if the product has latent defects that render it unfit for the use for which it was intended.

- It requires inherent defects that make it unfit for purpose or so impair its use that the buyer would not have acquired it, or would have given less for it, had he known of them.

For example, if a vendor sells you a watch with the assurance that you can dive with it to 3,900 meters like a Rolex Deepsea and it turns out that your watch has been damaged by water and pressure after your seaside activities, then it's a latent defect.

- The defect has to be hidden at the time of sale. Only an undetected defect brings the warranty into effect. As a buyer, you have to give the watch you wish to buy your full attention, checking that it works, that it is undamaged and in its own box with its accessories and certificate of authenticity. It is therefore up to you to check that the watch does not have any apparent defects. Furthermore, and especially for mail-order purchases, the defect must have existed before the transaction, with the onus of proof on the buyer, but the seller or even the judges are often understanding. Sales of second-hand goods have two particularities: the first is that ordinary wear and tear does not constitute a defect, and secondly that the buyer of a pre-owned watch should take special care to examine the object without being required to undertake detailed investigations.

- If the two conditions above apply to you, any action for latent defects must be filed within two years of the discovery of the defect.

Please note: Do not confuse the warranty against latent defects with the obligation of conformity.

The obligation to deliver is stated in article 1614 of the Civil Code: *"the thing must be delivered in the condition in which it is at the time of sale."* The object that is delivered must conform to the object sold and to the use for which it was intended. It is up to the buyer to prove any lack of conformance. Furthermore, the acceptance without reserve of a watch obviously lacking conformity will prevent you taking action for failure to deliver because the defects are apparent.

Le mensuel horloger

LA REVUE DES

MONTRES

N° 232 / FÉVRIER 2018 / MENSUEL

LAREVUEDESMONTRES.COM

RICHARD MILLE
RM 53-01

L 19083 - 232 - F: 3,00 € -RD

08:00 SPÉCIAL SIHH
Le salon qui recense les tendances et donne le top départ de l'année horlogère.

12:00 SPÉCIAL SIHH
Cartier, entre mouvement Mystérieux et rotations hypnotiques

16:30 SPÉCIAL SIHH
Montblanc : Star Legacy, deux pièces qui replacent leurs racines dans l'air du temps

M.M.C

MONTRES MODERNES & COLLECTION

www.montres-modernes.com

PATEK PHILIPPE
GENEVE
ROLEX
OYSTER
COSMOGRAPH
DAYTONA
UNITS PER HOUR
AUDEMARS PIGUET
AUTOMATIC
BREGUET
No 2649

ABP
CONCEPT

Atelier du Bracelet Parisien

TACHYMETRE
60
500
400
350
300
275
250
225
200
190
180
170
160
150
140
130
120
110
100
90
85
80
75
70
65
SWISS MADE

PREMIUMWATCH
MONTRES DE COLLECTION
DEPUIS 2007 PARIS

www.premiumwatch.fr

SUBMARINER
660ft = 200m
SUPERLATIVE CHRONOMETER
OFFICIALLY CERTIFIED
SWISS – T < 25

ROLEX
OYSTER PERPETUAL
SAB
FEB
OFFICIALLY CERTIFIED
CHRONOMETER
SWISS

ROLEX
GMT-MASTER
SUPERLATIVE CHRONOMETER
OFFICIALLY CERTIFIED
SWISS – T<25

Grégoire Rossier
Anthony Marquié
FLIGHTMASTER ONLY
watchprint.com
Grégoire Rossier
Anthony Marquié
MOONWATCH ONLY
watchprint.com
The Magic of Watches
Louis Nardin
GEORGE DANIELS
LA MONTRE
AUDEMARS PIGUET 20TH CENTURY COMPLICATED WRISTWATCHES //
Elizabeth Doerr
BRIDGING ART AND MECHANICS
THE UNABRIDGED STORY OF CORUM'S GOLDEN BRIDGE
watchprint.com
George Daniels
Un maître horloger et son art
BREGUET WATCHMAKERS SINCE 1775
LONGINES
LONGINES
LEGENDARY WATCHES